Fishery development experiences

'Struggle on'

Fishery development experiences

W H L Allsopp

Fishing News Books Ltd
Farnham · Surrey · England

The views expressed in this book are those of the author and do not necessarily represent the views of the Centre

British Library CIP Data

Allsopp, W. H. L.
Fishery development experiences
1. Fishery management
I. Title
333.95′6 SH328

ISBN 0 85238 134 4

Cover illustration shows sites of regional case studies

Published by
Fishing News Books Ltd
1 Long Garden Walk
Farnham, Surrey, England

Typeset by
Alresford Phototypesetting
Alresford, Hants

Printed in Great Britain by
Whitstable Litho Ltd
Whitstable, Kent

Contents

Illustrations

Foreword

When Dr. Allsopp and I first began working together in the decade of the 1960s, he was the innovative and dynamic FAO Regional Fisheries Officer for Africa and I was the recently-appointed Assistant Director General in charge of the Food and Agriculture Organization's world-wide fishery programs. Coming from a developing country himself, and working on the fresh water and marine fisheries development questions on the whole continent of Africa, Dr. Allsopp faced a task that was both challenging and extremely difficult. Everywhere one turned, African countries were newly achieving independence. Embryonic departments of fisheries often consisted of a single person, newly graduated and facing a myriad of questions, always with inadequate resources. The times were stimulating and exciting. Aid from the multinational organizations, particularly from the United Nations Development Program and the development banks, was more or less readily available. Bilateral donors were anxious to assist in the movement away from colonial status to functional independence. The catalog of problems needing solutions was tremendous and Dr. Allsopp took full advantage of the remarkable opportunities for helping to design aid projects and to transfer experience from a variety of sources to the emerging fisheries sectors of the emerging countries.

Though our paths have diverged in recent years, we have each retained our concern for fisheries development and remain actively involved. Now, 20 years down the road, it seems both feasible and vitally important to make an assessment of achievements, successes, failures, and to determine what lessons we may learn and apply for the future. The subject becomes particularly important during a period marked by worldwide recession and a tendency for development assistance to be curtailed.

A basic problem lies in the common property nature of marine resources and, until the recent worldwide extensions of jurisdiction, usually to 200 nautical miles, the unrestricted access of fishermen from many nations to the coastal fishery resources. However, the drawing of lines on the ocean has not as yet substantially changed the nature of many fishery development problems. Nations now free from competition from external sources are increasingly beset with problems of limiting their own fishing effort, capital investment, and pressure on the finite fisheries stocks arising from lack of control over their own fishing enterprises. The problems of the size and productivity of the fish stocks, of rational design for the supporting infrastructure in developing countries, of limiting entry, of building the structures of laws, institutions, administration, processing, transport, marketing, and of controlling relatively unrestricted national investment and entry remain with us.

In this setting, Dr. Allsopp has done a great service by carefully examining and analysing the roots of success or failure in a well-selected spectrum of fisheries development projects of various types and in various parts of the world. He has looked painstakingly at the elements in each project that have measured or limited its contribution to the rational development of fisheries, fresh water and marine, from aquaculture to distant water capture fisheries.

The recent worldwide extension of jurisdiction has awakened new interest in the potential for fisheries development and for increasing and preserving the contributions of vital animal protein that the vast area of the world surface covered by water can offer us. The annual world catch has levelled off in the last decade, underlining the importance of improved husbandry of the living resources of the seas. It is a pleasure to commend Dr. Allsopp's study and to urge every fisheries development planner to take careful account of his analysis of what has been done in this unique and vital sector of the world food supply system.

Roy I. Jackson

Introduction

In recent decades, the world has seen a great population crisis, food crises in different parts of the world, a global fuel and energy crisis, the recent economic crisis in the developed world, and the present liquidity crisis. The continuous unemployment crisis in the developing world is now supervening in the developed world. Developing countries have been so buffeted by flood, drought, and crop disasters that hardship and crises are normal. For many formerly self-sufficient developing countries, food has become in the 1980s what oil became in the 1970s. Food must therefore be garnered, conserved, and effectively used on a national scale particularly in developing countries. The food contributed by fisheries offers much opportunity for improvement, and success can be highly rewarding.

Although over 60% of the earth's surface is covered by water, less than 2% of our food comes from the seas. Since the 1950s, developing countries have made increasing efforts to develop their fisheries. Not only is fish a vital food for most of these Third World countries; fishing also provides employment and often export earnings. The world food crisis, increases in population, and inflation have largely nullified the efforts of such countries to improve living standards; fisheries therefore have become increasingly important for food and work.

According to estimates by the Food and Agriculture Organization, more than 25 million people work in artisanal or small-scale fisheries, which produce about half the fish eaten world-wide. The improvement of such fisheries presents a great challenge. Developing countries, scrambling to build up their fishing capabilities, cannot afford the burden of unsuccessful operations. Coastal states now have greater management responsibilities than before, as a result of the establishment of a 200-nautical-mile (361.8-km) exclusive economic zone (EEZ); thus countries need to collaborate more closely than before this new Law of the Sea to ensure fair and efficient harvesting and use of this renewable resource. The management of such fisheries is, comparatively, a recent development, and even the developed countries have not been able to conduct their fisheries without crises. To transfer the knowledge gleaned in managing fisheries from developed to developing countries has been less easy than with other natural resource industries, inasmuch as tropical fisheries differ radically from the fisheries of the temperate parts of the world where developed nations gained their experience; different types of management based on more detailed knowledge of the specific conditions are called for.

During recent decades, there has been in developing countries a gathering momentum of interest and hope for increases in fish production. The hope has not always been fulfilled. Indeed, the difficulties of managing fisheries have become more complex and numerous. Most marine fisheries have not been managed well; now there is a need for efficient management as well as for increased production.

During this period of increased expectations, a considerable proportion of investment in fisheries development in Third World countries has been from foreign rather than domestic sources. The investment thrust has come from international or multilateral sources, such as the United Nations Development Program through FAO, the World Bank, and regional banks of Asia and Latin America, and through many bilateral agencies such as the Canadian International Development Agency. Obviously, therefore, there would be value in reviewing, in depth, the experience of development projects that have been carried out, so that in Third World countries those who make development policy, their technical advisers, and those who carry out the projects can compare their own experiences with those of others doing similar work in fisheries.

Donor agencies customarily carry out internal assessment at the end of every project. However, such reviews are not widely circulated and are often available only to the agency's staff. Frequently the nationals of the recipient country who worked on the project do not have the benefit, for use in future projects, of access to the objective comments in these reviews. Development personnel working with other countries also may have no knowledge of such reviews, even though they may be dealing with exactly similar problems. It is obvious that sharing knowledge about what went right or wrong in managing various fisheries projects can be beneficial. It seems that personnel in government, industry, and commerce in other types of activity can have access to many case studies – but very few in fisheries.

The usual pattern in developing countries is that fisheries development starts with pilot projects, goes on to larger programs, and eventually results in industrial enterprises. When these developments, whether large or small, government or private, are successful, the direct beneficiaries tend not to publicize them, and therefore little about them appears in the literature. In fisheries enterprises the private sector prefers to hide its successes for competitive reasons; whereas in development projects government statements of success are often derided as political propaganda. Also, government development plans are often publicized as achievements too early, and public expectations are disappointed. Thus, one hears more often of failures than of successes in fisheries, and this regrettably contributes to a negative attitude at high policy level when patient encouragement is most needed.

Furthermore, who are the beneficiaries of a fisheries success? In small-scale fisheries, success usually means higher fish production and this, as will be explained later, may provide less benefit to the fisherman than to the market trader – who is the more skilled at turning production success into commercial success. Thus the artisanal sector continues to appear squalid, unhygienic, disorganized, and depressed despite considerable increases in fish landings. On the other hand, the introduction of mechanized vessels, and the establishment of small harbours or new installations convey to the public a more satisfactory impression of industrial progress, although such innovations are costly, demand greater management expertise, and are not necessarily as profitable.

In gathering together the information for this publication, it became obvious that the persons involved in the projects reviewed would offer candid, objective, and informative opinions only with some guarantee of anonymity. Accordingly it was mutually agreed not to identify the agencies that supported the projects or the countries in which they were carried out. The information thus made available is more comprehensive than would otherwise have been the case, and comment and criticism were less reserved.

The study was conducted through the courtesy of a number of organizations, chiefly the International Development Research Centre, of which I was a staff member on special leave. I received collaboration from the staffs of CIDA, FAO, UNDP, the International Fund for Agricultural Development, the World Bank, the Inter-American

Development Bank, the Asian Development Bank, the University of British Columbia (Institute of Animal Resource Ecology), the University of Hawaii (East–West Centre), Princeton University (Woodrow Wilson School of International Affairs), and various fisheries departments of African, Asian, and Latin American governments.

During the study I was able completely to review numerous documents pertaining to the projects concerned, as well as relevant regional literature. During site visits I consulted with project staff, fishermen, and others knowledgeable in the area's fishery problems, and I held final discussions with management and policymakers within each country.

To the many people who helped with information and discussion I express my gratitude. I have agreed to respect their confidentiality, but readers will appreciate the value of their candid assessments and comments.

2919 Eddystone Crescent
North Vancouver, B.C.
Canada

W. H. L. Allsopp

1
Historical perspectives and performance in fisheries development

Time frame for development

Among all the natural renewable resources, fisheries present perhaps the greatest challenge for effective management. The problems are the most difficult under tropical conditions. There are many interacting factors: a greater variety of fish species, changing climatic and oceanographic conditions (including massive seasonal outflows of rainfall from land), the level of exploitation by fishermen, the impact of other industrial activities, and the social, political and economic pressures on the market system. These factors all tend to be more extreme than in the older-established fisheries of the world, most of which are located in temperate zones.

In older countries, the fishing industries have evolved over centuries from small, individual, or family enterprises to corporate industrial, highly technical operations. In such fisheries the pace of development over the last 50 years has led to concern about the depletion of stocks, and this in turn has led to some international regulation of fishing; indeed management of fish stocks is now seen as vital for the economic survival of the fisheries. Scientists, operators, and policymakers now accept that safe levels of fish harvesting can only be assured by continuously collecting technical data about all aspects of a fishery.

It has taken decades of collecting and analysing data to assemble the necessary knowledge about the availability of species of fish, their behaviour, the effects of periodic climatic and oceanographic changes, and the expected trends in their population. Based on these data it has been possible to establish regulations, later to be improved, to ensure harvesting at safe levels in the long term. None the less, after centuries of fishing and with at least 50 years of concerted scientific and technical effort, fishery management is not an exact science, and it is not possible to predict complete success in the development of newer fisheries. Often the best that fishery managers can do is to apply the most appropriate scientific principles in interpreting and monitoring the statistical data collected from fishing operations when introducing upgraded technology.

It takes time to develop the capacity to manage fisheries. Adequate data for management decisions may be available only after 10 years of data collection. In the developed world, staff expertise in fisheries management has been built up slowly through experimentation, training, and experience, and this expertise cannot be transplanted to new types of fisheries, with unknown species of fish, instantly. Before government agencies can promote new equipment or production systems – such as gear, boats, and various other technical innovations – in an area, there need to be positive results from trials, possibly over several seasons. Staff from the developing country must be trained, often overseas, then given time to adapt under local conditions the principles they have learned. Administrations need time to develop confidence and build up support from their own government authorities; governments themselves need time to promote fisheries projects, which should be in relation to other programs and in collaboration with adjacent countries.

The foregoing is not an apology but a realistic appraisal of the time – perhaps 10 to 20 years – over which a government must commit itself to a fisheries program. Other natural resource development programs may take less time and accordingly look more attractive. However, short cuts in fisheries are not a realistic alternative; many programs start off, run into initial setbacks and are thereafter inhibited. Other fisheries programs achieve an apparent initial success that cannot be sustained.

The real issue is that governments feel they cannot afford the necessary preparatory period. Their need is so urgent that programs too often are undertaken without waiting for needed infrastructures to be ready; risks are taken and too often hopes remain unfulfilled. Any suggestion that the pace of development might be slower is often rejected as too conservative, or even too reminiscent of the era of colonialism. Progressive development is seen as quickly putting into operation large schemes. However, it has to be realized that the governments of developing countries suffer a dilemma of need: on the one hand they need optimum development of fisheries despite the limitations of their current services, which could take more than a decade to improve; on the other hand they need the fish, the jobs, and the foreign exchange *now*.

Experience has shown us that rapidly established programs are unlikely to lead to any lasting success. Deliberate and purposeful steps must be carefully planned for strategic modernization – that is, the development should proceed in logical sequence to the point at which its contribution to national life is optimal. And it is important to avoid project failures, not least because failures deter other fisheries development projects and cause investors, national banks, and private banks to put their limited capital into less risky ventures. If governments are not willing to persist until the fisheries management staff and infrastructures have been properly established, one cannot expect private entrepreneurial efforts to do any better.

Thus failure cannot be tolerated, yet the time available to prepare for an assured success is limited. It will be enlightening to see how fisheries projects in the Third World have met this challenge and what achievements there have been in actual programs. Through an analytical review we may also see whether any guidelines have emerged to ensure success in new fisheries programs.

Historical trends

During the past 30 years, many developing countries have made special efforts to develop their fisheries, because for most of them fish is a vital food supply, rather than being a luxury item. Faced after independence with increasing populations, these former colonial territories saw the potential of fisheries for helping toward national self-sufficiency in food and for providing employment. Accordingly they were impelled to promote the exploitation of these resources.

Unfortunately, however, development has been subject to problems – those of seasonal availability of fish, lack of technological expertise, shortage of economic resources, and jurisdictional aspects. The administrative expertise necessary to best develop and manage marine and inland fisheries has also been lacking.

Up to the mid-1960s, fisheries development activities were largely confined to the initial process of studying the nature of the resources available. Thereafter, there was increased emphasis on commercial harvesting. Governments thus learned by experience some of the realities of managing fish stocks, as well as the availability of the fish and the interaction among different species. The operations showed up deficiencies in staff and facilities. A difference of approach became evident – the Food and Agriculture Organization (FAO) and United

Nations Development Program (UNDP) emphasized studies of the rational development of the resource, although they did also provide technical assistance, directly and through regional commissions, for practical development activities. Governments, however, considered investment for production to be of higher priority than resource management programs.

In some regions, fish are often more abundant in the open seas – the areas outside what used to be territorial limits before the proclamation of the 200-mile (361.8 km) exclusive economic zone (EEZ). Such fish, however, derive from marine and inland waters, controlled by one or a number of nations. The fish stocks include sedentary species, such as shellfish, beside those that migrate across territorial boundaries and do not permanently inhabit the waters off any one nation's coasts. There are also seas, large lakes, and great rivers that are bordered by several countries, and this migratory feature applies there, too. Thus, the management of a fishery may not be wholly within the jurisdiction of one country but has to be undertaken jointly, perhaps as a regional effort. There is therefore a lack of definitive ownership of the fish stocks, which move freely in the oceans and cross national boundaries without restraint, a circumstance often referred to as the 'common property' nature of fish resources. Further, ownership of a fish only supervenes after it is caught by a fisherman, so common property pertains to the national level as well. The effective management of fisheries, given the common property nature of the fish and access to the fisheries of vessels from any nations, has not been easy anywhere in the world, but the problem is even more acute in new Third World fisheries, where the countries are attempting to develop their operations while simultaneously acquiring knowledge of the behaviour of the several species in their waters.

Recent programs

The total world fish catch has not increased much in landed tonnage since 1970 and was in 1981 estimated to be 72 million tonnes. During the past 20 years important fish stocks have, through increased fishing intensity and eventually overfishing, been brought to near-ruin. However, the introduction at the third Law of the Sea conference of the EEZ gives greater jurisdiction to coastal states over the fisheries off their shores. This offers the possibility of so managing the fisheries as to organize the fishing industry with rational catch limits. Thus investment in vessels and gear can earn a proper return, markets can be organized and satisfied, and the coastal communities can receive lasting benefits.

There have been many projects world-wide to promote the effective use of fisheries, and many of these projects have been assisted by United Nations and other multinational or bilateral agencies. The projects vary from providing individual advisers to supplying groups of technical experts; there have also been regional programs.

Many programs involved practical technology for improving the production, marketing, and processing of fish. Some assisted small-scale enterprise by testing a new type of vessel or a modification of a traditional craft, perhaps using a motor. Others introduced new designs of nets or other catching systems. Still others pioneered new shore facilities, such as fish-handling, storage, processing, packing, or distribution systems. While these activities were proceeding, some staff received on-the-job training and others were educated overseas.

Concurrently with these changes, the profitability of fisheries was further complicated by higher prices for fuel and the increasing need for food, jobs, and foreign exchange.

Development momentum

There has therefore been a world-wide acceleration of organized efforts to develop fisheries. Sometimes one success had led quickly to follow-up by communities or large private operators – this has happened after introduction of new fishing gear, processing systems, vessels, or motors. Any time an innovation captures the fancy of the fishing community – as for instance when outboard engines replaced paddles and sails on canoes – it has quickly been taken up by individual fishermen and given support by government credit systems. (The motorization trend caught fire and rapidly spread from country to country in West Africa.)

Often, however, the initial assessment of an innovation was not followed by a thorough evaluation of its long-term value. There have been few objective economic analyses of the full results of a program – its impact on a community or the limits to which it could be replicated on a wider scale, given available fish resources. Before such expanded development, data have not always been available, with the result that the exigencies of food and work have overridden the need for careful analysis that could have influenced the direction in which operations might go.

Initial efforts at development in one country stimulated major investments in similar countries through national, regional, and international banks. Competition – between the provinces of a country and between the countries of a region – to promote fisheries has developed. The programs thus stimulated have mostly been carried out by governments with technical help from aid agencies and financial help from international investment funds. However, the results have often been compared, unfavourably, with the results from other types of development investment. Yet quite a lot has been achieved in a relatively short time, considering the lack of infrastructure with which many projects began, and the inadequate preliminary data. Finally, in a practical way, governments have often found themselves with little real choice as to how to proceed.

Governments, for instance, often found that they were unable to stand back, restrict development, and conduct deliberate studies about a jointly owned resource which their neighbours were busily harvesting. Often joint ventures with foreign fishing companies seemed, in such situations, an attractive and prompt solution for private and state enterprise. Prudence was equated with the out-moded pace of development and governments proceeded to establish 'modern' facilities and operations before building up the necessary data-base and experience. Projects developed into programs, supported by various aid agencies, with varied results.

Production increase

From 1965 to 1970 the landed catch increased in many developing countries. So satisfactory did this seem that there were predictions that the total world production of fish would increase from 60 million tonnes to possibly 200 million tonnes. Later analysis showed that the pace of those early increases was, in fact, not sustainable. As a result of the decline in catches of Peruvian anchoveta and the levelling off of the Third World's catches, fish production has seemed to settle at around 72 million tonnes, of which 12 countries (Japan, USSR, China, USA, Chile, Peru, India, Norway, Korea, Denmark, Indonesia, and Thailand) account for 60%. It is now generally considered that the maximum sustainable yield from the world's fisheries will, by 1990, not exceed 100 million tonnes annually. Of this, an increasing quantity will be used for direct human consumption, much of it by the burgeoning populations of the Third World.

The artisanal fisheries' contribution to the total catch is highly significant in many countries, even though accurate statistics are not available.

The increase in available fish through successful investments, now and in the future, highlights the importance of developments in infrastructure that will enable its best use. This is a critical concern in Third World countries, where the necessary resources of money and expertise, as well as fish, have been in short supply.

Development climate

Fisheries in developing countries are being affected by economic constraints, a record of at least some disappointing investment experiences, and the challenges of managing the expanded area in the new EEZ. These problems mean that small countries need to develop their fisheries in phased strategic plans of modernization, but this has been difficult. None the less, they have been able to organize their fishery administrations, either by their own efforts, with the assistance of FAO regional fishery commissions, or through the UNDP regional development programs. It has become clear to governments that their national fisheries administrations need more trained staff, efficient supporting organizations, and appropriate policies to deal with social and industrial aspects of their fishing industries. They have become more aware of the scope and complexity of problems in marine fisheries for pelagic or demersal (surface- or bottom-dwelling) species, particularly through the collection of statistical data. They have come to recognize that some developments cannot be rushed and that the survival of a fishery often depends on effective monitoring for conservation.

The process of monitoring fisheries for conservation has not been markedly successful in the developed countries of the world; consider then the challenge that it presents to the developing nations, who aspire to manage more-complex tropical stocks of fish with little current knowledge of their behaviour and a scarcity of trained people. Obviously, the necessary enlightenment of Third World managers will take time but must be patiently and confidently pursued.

Performance

The success of Third World fisheries management has been varied. Some fisheries have progressed remarkably well; others have not, despite careful planning. Frequently the failures have been due to circumstances beyond the immediate control of fisheries management. In the developed world, efficient and successful ways of producing fish have been found; electronic and mechanical innovations from other industries have been applied to fishing, while advanced technologies of locating, catching, and processing fish have resulted in higher production per person involved, as well as new fish products for the consumer. This has also, however, resulted in overfishing, with the need for some restrictions in the well-developed fisheries. This has led to a dispersal of the fleets of developed nations to other, less-exploited fisheries of the world, often under joint-venture arrangements with Third World countries. Some developing countries also have had notable successes in increasing fish production, but the demand for fish in these countries – such as Peru, India, Thailand, Indonesia, Chile, Phillipines, and Korea – has not yet been adequately met. Some developing countries are better-informed than others about the technicalities of commercially important fish species, distribution, and migratory behaviour, but they remain unable fully to use their knowledge for lack of the ability to exploit fish resources.

Broad objectives

Fishery programs in developing countries have multifaceted objectives:

1. Contributing to the quality of life in rural and low-income communities through:
   - Helping toward national self-sufficiency in food by locally produced fish;
   - Increasing employment, either directly in fishing or in supporting activities;
   - Improving standards of health, education, and general prosperity in fishing communities.
2. Contributing to the national economy through:
   - Replacing imported food with local fish, thus saving foreign exchange;
   - Processing fish for export, thus earning foreign exchange;
   - Stimulating national investment in fisheries.
3. Maintaining the fisheries on a permanent basis through:
   - Management of the fisheries to ensure sustained fish harvests at an optimum level, avoiding over-investment;

- Enhancing the quality of management to enable it to collaborate effectively with other countries that harvest the same resource;
- Conserving fish stocks against over-exploitation and protecting water resources against degradation from other competitive use.

Developing a fishery

Any national development plan for fisheries may aim at a number of the above objectives. Developing a fishery calls for integrated action at several levels. Personnel will be engaged in:

- Policymaking at the central and local government level;
- Administration;
- Technical management, including research and development;
- Private activities, including harvesting, processing, marketing, distribution and investment;
- Operating public facilities, such as harbours, transport and markets;
- Public sector services, such as extension, training, and credit;
- Liaison with consumers, including researching their preferences, purchasing power, and total demand.

Those who develop a fishery must also, of course, collaborate with their opposite numbers in adjacent countries that share management of the same resource of migratory fish. This is a responsibility that has grown, particularly for small countries, since promulgation of the EEZ. As previously explained, these unprecedented problems have not been fully overcome even by the experienced staffs of the developed world's fisheries; accordingly personnel in recently emergent countries with still more complex problems bear a heavy burden. Despite the limitations of their training and experience, they have to meet the challenge to find urgent solutions to problems of food, unemployment, and even survival of some communities.

Governments can ill afford to wait for the evolutionary development of their fisheries sectors. Impatient as they are, they see neighbouring countries, through joint-ventures or otherwise, preparing to harvest a greater share of their traditional resource. And they feel they need demonstrable success or tangible installations or loan schemes to convince their fishermen and the whole country they are making progress. Thus, political expediency causes them to act before they have the needed data and staff or before the action may be economically justifiable.

Larger investment

From artisanal fisheries, the sequence of development next proceeds to larger investments financed by national development banks, commercial banks, regional investment banks, or large interregional or multinational banking agencies. The announcement that a project is approved by one of the large funding agencies often stimulates a ripple of confidence in local investment. Banks, credit agencies, and even moneylenders tend to provide loans promptly, in competition with the national bank. This tendency can even result in over-investment, thus putting into service more vessels and processing plants than are needed for the sustainable harvest from the resource.

Industrial fishing ventures may also be set up or improved by entrepreneurs, who operate without government involvement and may gain business expertise by some form of collaboration with international companies.

International aid projects are important in stimulating this type of development. However, this study does not review instances in which bilateral aid has supported industrial-type developments, but concentrates on artisanal fisheries projects. Artisanal fishing is the area of greatest concern in most developing countries, because it has the greatest impact on employment and fishing communities.

Regional programs

Regional programs form another series of development projects. Where a number of countries share a fishery resource, effective management clearly can best be carried out by all users, simply because any one national conservation measure may have little overall impact, whereas any one national action that is harmful can affect the whole resource. The critical need, then, is for collaboration, and the best examples of this are the regional programs sponsored by the global and inter-regional division of UNDP. These are illustrated in this book by the South China Seas Development and Coordinating Program and the East Central Atlantic Fisheries Development Program (chapters eight and nine respectively). Both were started with the help of UNDP and managed by FAO. The former project, after 11 years, is now completed, and the latter remains in operation and will continue for a few years yet. Both had significant support from bilateral aid agencies. Both had to deal with complex management problems. They are, however, excellent examples of the basic procedures that are necessary to accelerate regional fisheries – sensitizing government personnel to the issues, analysing data, and training personnel while catalysing technical improvements. The studies that were part of these two projects helped countries arrive at more enlightened policies for fishery development, investment, legislation, and trade.

Thus these two regional programs are important as case history studies for small, developing nations. Chapters eight and nine review the programs, in particular the input of a bilateral aid agency (CIDA) into a multilateral aid project. This combination of bilateral and multilateral aid may be a logical pattern in future projects; that is, a pre-investment study may lead on the one hand to regional cooperation with individual national investments; on the other hand bilateral aid programs may come in under this UNDP umbrella. The pattern seems particularly valid where many small countries share a resource; where there are various levels of development with a sizable artisanal component; where there is promise for industrial operations but insufficient supporting infrastructure; where human and financial resources are insufficient; or where the distribution of the catch among coastal states is disproportionate. It is also logical for bilateral inputs from countries with appropriate expertise to be incorporated under such an umbrella UN/FAO program.

Post-facta diagnoses

There have been some notable successes in fisheries investment programs in developing countries, but the record of performance is not wholly favourable. It was therefore considered desirable to review a series of case histories to assess the performance of projects that have been completed. These projects were either complete in themselves or parts of ongoing programs.

The results have been compared with the objectives set during appraisal and implementation, and the situation after completion of each project in relation to the inputs has been examined. Hindsight shows us what there is to learn from these projects that can be applied to future activities. It may be seen how recipient and donor can refine, improve, and speed up the technology transfer and delivery process. The reviews also look at what are the realistic schedules for a project and how project managers can cope with changing circumstances, such as internal changes in the country, new regulations in the fisheries, economic trends, climatic changes, social developments, and new factors that affect local labour. External factors with which project managers may have to cope include the availability of goods and services, trade developments in other countries, fishing activities by other countries or regional fishing developments, and changes in the migratory patterns of fish.

These factors are all considered in these reviews, which show the need for flexibility in carrying out a project, while keeping the aims clearly in view and using all practical ways to ensure success.

Reference has been made in the introduction to the secretive nature of those engaged in fishing, which shows up in the reluctance to provide details of operational data. This also has the effect of giving greater emphasis to the negative aspects of projects, sometimes making them appear less successful than they really are, so as to discourage competition from new entrants. So it becomes necessary to look deeper into these projects and to look at the lessons available from more developed, older-established fisheries. From these we see that successful enterprises have to be nurtured through a period of several, variable fishing seasons, with effective management, before they can be sufficiently established to cope with the hard times that are cyclical in fishing.

Evaluation procedure

The many agencies involved in development have now accumulated considerable experience in assessing the results of projects they have supported. Several books (see references at the end of this chapter) have been written on evaluation. In this book, a simple but comprehensive format, one pertinent to fishing, was sought for the reviews. After consulting systems used by ADB, CIDA, FAO, IDB, IFAD, IDRC, UNDP, and WB/IBRD a format was devised. The terminology adopted was that widely accepted and understood by the Interagency Task Force on Rural Development of the United Nations Advisory Committee for Coordination.

The case histories first describe the fishery situation in general and give the background and objectives of the project. This is followed by:

- *Project inputs*: the means provided by the project to achieve its objectives, including capital, manpower, equipment, and support services.
- *Project outputs*: the outcome of the project, such as installations built, trainees, items, and supplies and services generated as the result of project inputs.

- *Project effects*: the outcome or consequent use of project outputs – increases in yields, compared with the objectives and generally evident as positive or negative trends after the project is complete.
- *Project impact*: the continuing changes, attributable to the project, among the beneficiaries and the community in general, mostly near or after the end of the project. This includes better living standards, nutrition, incomes, trade, community participation, self-reliance, and policy changes. It also includes the influence of fish resources attributable to the project; project impact is not necessarily confined to the locality of the project.
- *Evaluation and assessment*: This is an analysis of the information available to show whether the project delivered its outputs and generated the desired impact. Thus, project achievements, difficulties, and operations are described, management is reviewed, and the economic returns, follow-up, and overall impact in the region are examined. The lessons learnt as a result of the project are also identified.

Selected references

Birgegard, L.-E. 1980. Manual for the analysis of rural underdevelopment. Upsala, Sweden, Swedish University of Agricultural Science International Rural Centre. 149 p.

Campelman, G. 1976. Manual on the identification and preparation of fishery investment projects. Rome, Italy, FAO. Fishery Tech Paper F111/T149. 86 p.

CIDA (Canadian International Development Agency). 1979. Manual sur l'évaluation de projects: directives, méthodologiques et différents aspects du système d'évaluation de l'Agence. Ottawa, Canada, CIDA. 127 p.

1980. A methodology guide for project teams responsible for managing evaluations. Ottawa, Canada, CIDA. 106 p plus 6 appendices.

FAO (Food and Agriculture Organization). 1979. Evaluation of technical cooperation projects. Rome, Italy, FAO Evaluation Service. 12 p.

IADB (Inter-American Development Bank). 1978. Guide for the preparation of fishery loan applications. Washington, D.C., USA, IADB Project Analysis Department. 53 p.

IBRD (International Bank for Reconstruction and Development). 1974. General conditions applicable to loan and guarantee agreements. Washington, D.C., USA, World Bank. 18 p.

1979a. Agriculture project analysis case studies and exercises, Volume 1. Problems. Washington, D.C., USA, World Bank Economic Development Institute. 710 p.

1979b. Agriculture project analysis case studies and exercises, Volume 2. Leader's guides and scripts. Washington, D.C., USA, World Bank Economic Development Institute. 113 p.

1979c. Agriculture project analysis case studies and exercises, Volume 3. Solutions. Washington, D.C., USA, World Bank Economic Development Institute. 157 p.

IDRC (International Development Research Institute). 1973. The benefits of project evaluation to IDRC and AFNS. Mimeograph. Ottawa, Canada, IDRC. 20 p plus annexes.

1974. Benefits of project evaluation: revisited. Mimeograph. Ottawa, Canada, IDRC. 11 p.

IFAD (International Fund for Agricultural Development). 1979. Operational guidelines on monitoring and evaluation of agricultural projects. Rome, Italy, IFAD. 56 p.

Price-Gittinger, J. 1972. Economic analysis of agricultural projects. Washington, D.C., USA, World Bank for E.D. Institute of IBRD. 221 p.

Sohm, E. D. M. B. 1979. Initial guidelines for internal evaluation systems of U.N. organizations. Geneva, Switzerland, U.N. Joint Inspection Unit. 43 p.

Squire, L., van der Tak, H. 1981. Economic analysis of projects (a World Bank publication). Baltimore, USA, Johns Hopkins University Press. 153 p.

Treasury Board of Canada. 1981a. Guide on the program evaluation function. Ottawa, Canada, Department of Supply and Services. 86 p.

1981b. Principles for the evaluation of programs by the federal departments and agencies. Ottawa, Canada, Department of Supply and Services. 46 p.

UK Foreign and Commonwealth Office. 1973. Guide to project appraisal in developing countries. London, England, ODA. 136 p.

UNIDO (United Nations Industrial Development Organization). 1980. Manual for evaluation of industrial projects (joint publication of UNIDO and the Industrial Development Centre for Arab States). Vienna, Austria, UNIDO ID/244. 139 p.

US AID (United States Agency for International Development). 1975. Guidelines for evaluation of capital projects. Washington, D.C., USA, Department of State. 32 p.

2
Artisanal fisheries infrastructure services – Middle East

Project background

The first of our case studies is a loan project aimed at increasing by about 6% the catch of coastal fishermen in a small middle-eastern country. This country has few known natural resources and before independence depended largely on servicing Suez Canal traffic. To diversify the economy, the government decided to develop the food-producing sector, especially fisheries. Financial, technical, and managerial resources were sparse, and most public investment was financed with foreign assistance.

Fisheries offered the best prospect for development. With a coastline of 1 000 km the potential yield of fish was more than 300 000 tonnes, whereas about one third of that was then being landed. Marketing arrangements were poor; although coastal residents consumed fish as a large part of their diet, people living inland did not, because of poor distribution, price fluctuations, and seasonal variations in supply.

The fisheries contributed less than 3% of the country's gross domestic product. The country had a population of less than 1.5 million, of whom 8 000 were fishermen, living in about 30 fishing villages. The fishing boats were houris (up to seven metres long, powered by sail and paddle) and sambuks (11 to 13 metres, sometimes with auxiliary engines). Seventy percent of the catch of the houris was sardine, used mainly for fertilizer and animal feed. About 15 000 tonnes of fish, including tuna, queenfish, mackerel, and lobster, were exported.

Before undertaking the project reviewed here, there had been joint ventures with two major fishing nations. The government had set up a corporation to deal with large-scale industrial fishing, provide credit to fishermen and undertake training and research. Fishery cooperatives were being established throughout the fishing community, with grants for gear and boats available through the government corporation.

Recognizing the shortage of skilled middle management and expertise in international fish marketing, the country sought bilateral aid to provide training for marine industry operatives and researchers to study oceanography and fish resources. The government subsequently recognized the importance of fishing by upgrading the corporation to a ministry.

Objectives

The project was intended to improve the traditional coastal fisheries by enabling the coastal catch to increase by about 75 000 tonnes a year, mostly tuna, queenfish, and mackerel for processing and export. The sponsor of the project, an international lending agency, would supply credits in foreign currency of $3.5 million to finance the external costs of the project, identify a second fisheries development project, and prepare a watersupply project. Local costs to be borne by the government were to be $685 000.

Tangible benefits from the project were to include:

- New shore facilities comprising fish landing sites, icemaking, processing, freezing, and cold storage plants, and receiving stations.
- Three five-tonne insulated trucks.
- New radio equipment.
- Five 14 metre demonstration vessels with engines, gear, and modern mechanical line haulers.
- Ninety-five traditional sambuks with inboard engines but with traditional manual gear and iceboxes.

The project was scheduled to start in 1973 and continue for five years. The economic rate of return was assessed at 30%. Technical assistance comprised a project manager anu two engineers with consultants for the organizational structures, harbour study and engineering, assessing the fisheries resource, and the watersupply scheme. Procurement for the supply of the gear, engines, and shore installations was through international competitive bidding, and the hulls were to be obtained locally.

The estimated 30% rate of return plus the increase in local food supplies and foreign exchange earnings amply justified the project. Better boats would produce better incomes for those fishermen who took part in the project, and the increase in fishing capacity would also provide extra employment in the project area.

An assessment revealed that resources were adequate, that coastal

Fig 1 Traditional Dhows or Sambuks fish the Red Sea

fisheries had development potential, and that as deep-sea resources were already being exploited by joint ventures it was the less-prosperous artisanal sector that could produce the best results in increased fish, jobs, and foreign exchange.

The assessment showed inadequacies in the administration of the public services necessary to support the project. It was, however, assumed that middle management and lower-grade personnel would be available to run the shore facilities. This was an optimistic assessment, perhaps because the project was one of the first to deal with fisheries by the agency and was also one of the first such projects for the host government.

Delays in appointing a project manager and consultants, in disbursing loans, and in setting up a department in the host government to handle the project combined to defer the start of the project by a year and a half to 1974. At that time a reappraisal put the cost at $6.2 million instead of the $4.2 million first envisaged. (By the end of the project the host government had increased its contribution by $1 million and the agency had increased its contribution by $2.4 million.)

Project inputs

Four missions carried out identification of need, pre-preparation, and appraisal before the project began, accounting for 55 person-weeks. There were 12 supervisory missions and, before the end of the project in 1981, two completion missions.

The slow recruitment of a project manager and engineers resulted in a delayed start and implementation of the project, causing a 56% escalation in costs. The first deliveries of boats occurred before the harbour installations were ready. Cost-cutting by reducing activities and re-deployment of boats to other areas, thus limiting the number of boats for supplying the shore processing facilities, adversely affected viability of the project.

Even after construction of the shore installations began, progress was slow. Local construction companies found skilled labour hard to get. The expatriate consulting engineering company had insufficient overseas experience and went into liquidation midway through the project; the project manager hired individual engineers to finish the job. The construction work turned out to be a learning experience, which increased local capabilities, but this gain was at the expense of speed and economy of construction. The job took twice as long to do and cost 38% more than had been planned.

After the shore installations were built, the lack of trained manpower continued to make itself felt. It had been planned that there would be expatriate technical aid to give on-the-job training to local staff. However, national staff from the bilateral project was trained in other disciplines. The supervisory and completion missions found a continuing shortage of local expertise. The result of this was that the shore facilities were used to only 20% of their capacity. Production did not attain its targets for several other reasons, one being the difficulty of trucking frozen produce inland except in small quantities.

The sambuks also took longer to build than had been planned, because of the shortage of skilled labour and materials, but when put into service they succeeded in catching more fish than originally estimated. However, although the 100 boats were built as originally intended, they were distributed along the coast; thus the catch landed in the vicinity of the plant was insufficient to make it economically viable. The plant therefore also accepted catches from offshore trawlers.

Despite these operational difficulties, the project did succeed in bringing into being an administration that was capable of managing fisheries development projects, and that in fact later did so successfully. However, the administration of the loans to local fishermen for fishing vessels, through their cooperatives, was not effective.

During the implementation of the project, the government set up a department to deal with projects of this type, thus enhancing its ability to supervise development. Gradually, the government acquired the determination and the machinery to develop fisheries, including credit management. Subsequently was an intermittent review of fish prices paid by the government corporation. Although these revisions may not have been based on examination of all factors – local market prices, export prices, handling costs, etc. – the innovation was a good start.

Outputs

The cold storage installation and harbour, although behind schedule, were eventually completed. Lack of competent management produced difficulties in full operation; a management study suggested changes in organization, structure, and accounting methods, and these were eventually adopted. A harbour development study suggested the plans be modified and smaller installations than originally planned were built, and fish production from all vessels was then processed. The numbers of personnel trained and the level of credit services were inadequate, but later projects were planned to remedy these defects.

Sharp increases in the price of imported timber increased the price of the boats from the originally projected $6 310 to $19 545. Timber-supply problems also led to delays in production.

Effects

The original plan of using five of the 100 boats as demonstration vessels was shelved. The 100 boats that were built were distributed to different regions along the coast, a decision based on political considerations. The boats operated by fishermen and owned by cooperatives consistently produced catches one third higher than expectations, thanks to an incentive scheme. Government agency-run boats did not reach their production targets, and overall the increase in catch did not reach projections. Subsequently the government nationalized all private sambuks.

Before the project began, the appraisal suggested little difference in benefit between importing wood to build boats and purchasing glass-reinforced plastic boats abroad. However, as the project proceeded, the cost of wood rose. Subsequent experience in neighbouring countries has shown it is practical to manufacture glass-reinforced plastic boats locally.

The ability of local construction companies and government services to undertake the fisheries installations was inadequate. However, the government insisted that local contractors get the work. They had very limited experience with this type of construction and so had to rely entirely on the foreign consulting engineering firm, which went into liquidation during the project.

There was no provision for training personnel to operate these facilities, when built, apart from some on-the-job training from technical assistance personnel, who had other demanding duties. The investment agency acknowledged that the training should have preceded the construction. At the time of appraisal, the agency conceded, it had believed that infrastructure rather than trained manpower was the critical need. It was only during the implementation of the project that the manpower problem became fully apparent; emphasis in subsequent projects therefore shifted to training and technical assistance. At the appraisal of the loan project it was assumed that national support staff would be available, as a bilateral aid agency had been providing equipment and training for a fisheries and oceanographic project. However, neither the personnel nor the equipment was suitable for use in the project reviewed here, nor was it available for this project. In short, initially (in 1970) it was assumed that the personnel trained "for fisheries" would be suitable managers, but it was later (1975) realized that their training was theoretical and scientific and not for the practical fish plant operational management required by this fisheries project.

The system of providing credit that was introduced proved to be unsuitable and needed to be modified with the services of staff trained in administration of loans. The monitoring of monthly catch data of fishing operations was important for evaluating the project.

The government agreed to undertake this responsibility but did not have the staff to carry it out. Indeed, national management as a whole was inadequate to operate such a large, new scheme.

Impact

The socioeconomic benefits of the project included increased catches and higher incomes for fishermen. Carpenters and other workers at the boatyards and shore facilities obtained employment. Better levels of nutrition seem to have resulted from the greater fish catches. The interest

Fig 2 Boatyards motorize and modify designs

of local people, notably in their cooperatives, led to greater self-reliance that augurs well for further development; indeed the government became enthusiastic over two more projects, which subsequently got underway.

Evaluation and assessment

Both the government and the investment agency emphasized in this project improvement of infrastructure and fleet development. These improvements outpaced the country's capacity to provide trained and qualified people to manage these investments. Until such people were available, the investment itself was precarious. By contrast, in the industrial sector of the fishery, expatriates largely managed the joint ventures to protect the investment of foreign partners.

The government and the lending agency, in the project under review, did not, at the outset, seem fully to appreciate the extent of management skills and staff support required to undertake the complex and integrated activities of production, handling, processing, storage, and marketing of the fish – a very perishable commodity. Because the shore facilities were not well managed, the quality of fish produced was poor and much of it spoiled.

During the course of this project the need for manpower training was seen, and accordingly a subsequent project was organized. It might better have been an integral part of the original program. Technical assistance to alleviate manpower deficiencies is demonstrably of a higher priority than extra physical infrastructure – in this country investments worth $200 million have been made in trawlers, cold storage, canning, processing, and other shore facilities. What was needed was management inputs (investment in skills) to enable them to operate more effectively.

The equipment supplied was adequate, although its delivery was slower than expected. However, eventually the new gillnets, sharklines, and other improved gear were put into service and catches increased.

Government was slow to set up support services at first, but eventually this process gathered momentum toward the end of the project and in the follow-up projects.

In its self-assessment, the agency concluded that there seemed to have been over-optimistic assumptions about the time the project would take to get under way and the efficiency of local operations. The economic benefits forecast did not materialize to the extent hoped for, and at the end of the project the whole scheme was not financially satisfactory in the recovery of sub-loans to purchasers of boats or in the throughput of the shore plant.

Eventually, after a fish-sales price review in 1979, with a doubling of the 1977 prices, the liquidity of the fishermen's cooperatives improved.

The appraisal was made with a limited knowledge of the potentials and constraints of the country. This led to a poor project design and poor timing of the sequence of activities. The project did make a major contribution to fisheries development. It stimulated improvements in productive capacity and administrative ability. Those who worked in the project learned more about the country's fishing potential and limitations, and this experience proved valuable for further development and the operation of subsequent projects.

The management authority had not acquired adequate experience to operate the cold storage facilities responsibly and effectively. The government corrected the shortcomings of the shore facilities after visiting review missions reported adversely on conduct of operations.

This was one of the first such investment projects of an external investment bank, so it is easy to understand, in retrospect, that the difficulties now recognized to be inherent in developing artisanal fisheries should not have been appreciated. This type of project calls for more than just new buildings. It is a long-range development of implanting an infrastructure and gaining experience and community confidence so as to develop the social structure and community activities of artisanal fisheries.

The lending agency seemed to have capitalized the project in a way that made it characteristically more of an instant industrial operation. However, the business and operational expertise characteristic of industrial-type fisheries has to be evolved gradually in artisanal fishing communities. The operation of shore installations by nationals with adequate experience required more adequate training locally and overseas. Without such training there is no possibility of lasting impact and forward planning in the operation and management of processing facilities. Breakdowns and bottlenecks that plague artisanal operations will prevail. There, stoppage is less financially critical because of less capitalization than in an industrial enterprise.

Adequate training requirements for local management staff seem to have been completely overlooked. Perhaps it was assumed, at appraisal, that bilateral programs would produce adequate numbers of trained nationals, but it must soon have become evident that appropriate specialization of trained staff would be needed at the beginning of the program if it were to maintain the schedule set. Such staff requirements must be completely catered for within the project itself. If necessary those who organize the project must ensure the training of personnel before operations commence so that the necessary foremen, support personnel, and counterparts to the expatriate advisers and experts will be in place. Evidently one cannot depend for this on bilateral inputs or local sources of training, as these may not be relevant, adequate, or appropriate for the new organization that has to be created.

Choice of the project manager was critical to the success of this project. Indeed, his skilful management avoided the collapse of the program and established an excellent rapport with policymakers and local executives, who subsequently supported other projects. The confidence in him was such that he was able to proceed without interference in fish-plant and project operating decisions.

The program as approved at appraisal seems not to have been flexible enough. Apart from the project's initial failure to cope with the training problem, it seemed unable to have accommodated local changes such as in the allocation of vessels, fish pricing, loan policies, and legal commitments. Flexibility might have been improved had there been provision for a midterm review. Further, a system of continuous review with the government through establishing an appropriate management committee could have spotlighted difficulties to both parties and perhaps mitigated some of them. Nor did the program adequately take account of changes in the catch or social and economic changes taking place in the community.

The rate of return of 30 – 35% was over-optimistic. It might have been more realistic to predict lower catches according to known performance and accept a lower return. It might have been wise to recognize the probability of delays in deliveries and implementation, although this is a factor more easily seen in retrospect. Perhaps the earlier optimistic projections were believed to be in line with what the investment agency, accustomed to enterprises more commercial than artisanal fishing, would expect to be told. Rates of return are more an indication of feasibility than an end in themselves. Markets and prices were disappointing in that at one stage the government was slow to permit payment of incentive prices to fishermen; when this changed the prospects were better.

In view of the considerable investment with which counterpart staff was entrusted, there needs to be in such programs a clear prior understanding of the technical capability and training they should have. This should be embodied in the conditions of the loan. To do so is in the interests both of lender and of host government.

The project planned too great an input of infrastructure all at once, especially in view of the limitations of personnel. Development in stages would have been better. A massive physical plant with limited management, or a large number of boats with a consequent heavy demand for maintenance operation and product handling, can obviously create severe problems.

The frequent supervisory visits were valuable. However, had an advisory or review committee existed, it could have met the visitors, along with the project manager and staff. It would have explained autonomous national decisions and perhaps gained support for them; likewise it could have reported back with advice to the policymakers. It seems that the supervisory visitors were unable to hold collective reviews and met only with individual executives. Such a local steering committee would also help with credit applications, loan recovery, pricing policy, the uses to which project outputs might be put, and supporting services.

The choice of project manager may have been the most critical factor in the success of the project, but once he has completed his work and left, it is the national program co-manager and his support staff on whom lasting impact will depend. For this project, a developer and organizer type of technician was likely to be a better choice as co-manager than a highly trained, report-writing, research-oriented person; personnel-management, or business experience was the capacity required. Managers in the type of program described here need on-the-job experience in fishery plant and fleet operations. It may be necessary to use a national manager with experience in related industries until persons with a background in fisheries can be trained in management. The adage that 'in developing countries the best investment is the training in management and skills' clearly applies to investment projects like this.

Overall impact on country

The most effective operation of the project was the improved organization for building fishing vessels. The experience of building improved wooden sambuks had a lasting effect on the industry, even though the high cost and scarcity of imported teak made it necessary to switch to building smaller, glass-reinforced plastic boats.

In retrospect it is easier to see the need for establishing an independent local body to monitor and evaluate the activities of the project and advise on loan policy.

The most difficult activity was the construction of installations owing to staff limitations and the lack of skilled masons, carpenters, and other tradesmen. It would have been useful to have had an advisory committee to discuss staff constraints at the beginning of the project and to have had trained nationals available at start-up. The lesson has been learned for future projects.

The cold storage has functioned usefully to stabilize the market. It now caters to artisanal production and takes the regular catch from joint-venture industrial fishing. The operational experience has assisted the orientation of subsequent projects in the country and region.

Selected references

Abbes, R. 1978. Suggestions pour l'établissement d'un Plan de développement de la Pèche â Djibouti. Rome, Italy, FAO-IOP. 17 p.

Anon. 1981. Special review: an integrated approach to the development of fisheries in the Northeast region. Agriculture and Development, 4, 13–18.

Barraniya, A. A. 1979. Technical report on the socioeconomic aspects of Red Sea fisheries. Rome, Italy, FAO-UNDP. 57 p.

Barraniya, A. A., El Shennawi, M. A. 1979a. Summary report of survey of Egyptian Red Sea fisheries. Rome, Italy, FAO-UNDP. 60 p.

1979b. Exploratory socio-economic survey of Sudanese Red Sea fisheries. Rome, Italy, FAO-UNDP. 36 p.

Bromiley, P. S. 1972. Marketing of Red Sea demersal fish. Rome, Italy, FAO, IOFC/Dev/72/26. 23 p.

Campleman, G. *et al* 1977. Fisheries and marketing in the Yemen Arab Republic. Rome, Italy, FAO, YEM/10/TEM, IOP/TECH/77/13. 85 p.

Cox, K., di Palma, S. 1978. Small scale fisheries development in Djibouti. Washington, D.C., USA, USAID.

FAO (Food and Agriculture Organization). 1976. Report of the UNDP/FAO review and evaluation mission of the IOF Development Programme. Rome, Italy, FAO, Indian Ocean Fishery Survey and Development Programme, 19 January to 18 February. 61 p.

1977. Round table discussions on regional approach to fishery development, Svanøybukt, Norway, 30 August to 4 September. 1976. Rome, Italy, FAO IOFC/DEV/77/42. 18 p.

1978. Report of the steering committee for the development of fisheries in areas of the Red Sea and the Gulf of Aden. Rome, Italy, FAO M/S Travel Report. 57 p.

Undated. Fishery country profiles for Djibouti, Egypt, Ethiopia, Somalia, Sudan, Yemen Arab Republic, People's Democratic Republic of Yemen. Rome, Italy, FAO.

Guckian, W. H. *et al* Reconnaissance survey of fishing vessel construction and repair facilities. Rome, Italy, FAO, IOP FAO-UNDP IOFC/DEV/71/10. 31 p.

IFAD (International Fund for Agricultural Development). 1980. Artisanal fisheries pilot development project, Djibouti. Rome, Italy, Arab Fund for Economic and Social Development/IFAD. 34 p plus annexes.

Moal, R. 1977. Perspectives de développement des pêches en République de Djibouti. Paris, France, SCET. 23 p.

Mostrup-Schultz V., Sorensen, R. 1979. Development of landing places and minor fishing harbours in the Yemen Arab Republic. Rome, Italy, FAO. TRAM M/S. 10 p.

Pajot, G. 1978. Report of the Indian Ocean programme/FAO fisheries mission to Djibouti. Rome, Italy, FAO, UNDP/F10/IOP/TRAM/1484. 13 p.

Payne, R. L., van Santen, G. 1977. Report of a fisheries mission to Djibouti. Rome, Italy, FAO, UNDP/F10/IOP/TRAM/1390. 4 p.

Silva, L. I. J. *et al.* 1978. Development of fisheries in the exclusive economic zone of Somalia: report of an FAO/UNDP international Indian Ocean fishery survey and development programme mission. Rome, Italy, FAO, IOP/TECH/78/19. 32 p.

3
Motorization of a West African canoe fishery

Background

In this chapter a project to increase the number of motorized canoes in a West African country is reviewed. The national economy was based on agriculture, but the country had developed its fisheries to make an important contribution to coastal employment and food supplies. Annual production of fish exceeded 300 000 tonnes, of which about 180 000 tonnes came from the artisanal sector. About 27 000 fishermen operated more than 5 000 canoes, of which roughly half used sail and paddle and half had small outboard motors. The coastal population had a well-established seafaring tradition, and fish seemed abundant. The catch consisted of 60% pelagic fish, 20 to 25% demersal species, and the balance estuarine fish and shrimp. The industrial fisheries produced about 80 000 tonnes of tuna, from longlining and purse-seining, to supply canneries and fishmeal plants.

The artisanal fisheries provided year-round employment for coastal communities, where handlines, encircling gillnets, beach seines, and bottom gillnets were used seasonally to catch fish, shrimp, and lobster. Eighty-three percent of the catch was marketed fresh. Annual per capita consumption of fish throughout the country was 28 kg, against 9 to 20 kg in Europe, for instance. Coastal centres processed fish in various traditional ways for distribution to markets throughout the country.

Artisanal fisheries had already been given high priority in development plans before the start of the project described here. There had been a previous motorization project that had increased the proportion of motorized canoes to nearly half.

Rationale

Motorization would enable canoes to spend less time getting to and from the fishing grounds, make longer trips with more gear and less crew, while still reducing the physical strain on crews, and land their catches quicker, thus producing a higher quality of fish. The increased catches would improve earnings and enable fishermen to achieve a higher standard of living. However, the operational system would have to be well organized so that fishermen did not have problems with maintenance and would be able to keep their boats in service.

The cost to the fishermen of motorization should be as low as possible. Initially, expatriates would organize the maintenance services and operate the storage and marketing centres, these functions to be taken over in time by local staff. An efficient maintenance and spare parts supply service would provide an incentive to the fishermen to join their local fishermen cooperative associations. A central parts depot-cum-service training centre was to be established at the project headquarters and workshops, while supply stations were to be set up at seven key locations along the coast. Arrangements included centralized technical services, inventories for the supply of gear at low cost, and spare parts at the major fishing centres. Motor transport and radio-telephones would link headquarters with the outlying centres.

Objectives

The aim of the project considered here, then, was to complete the motorization of the canoe fishing fleet and set up coastal supply and

Fig 3 Propelled with outboard engines dug-out canoes are first on the fishing grounds throughout West Africa

Fig 4 At the landing beach the engine is taken ashore

service centres. Experience had shown that motorization increased the average catch per unit by 60%, from 22 to 35 tonnes a year. The program therefore aimed at further increasing the catch. Another objective was that fishermen's cooperatives should manage fishing centres.

The project planned to supply 3 500 outboard engines, with adequate spare parts, equipment for maintenance workshops, and cold storage installations. There would also be technical assistance services and local and overseas management training of nationals. The projected outlay was to be $1.9 million for the motors, $567 000 for spare parts, $253 000 for tools and repair workshops, and $390 000 for personnel services, for a total of $3.1 million. These funds were to come from a bilateral donor, partly as a loan, partly as a grant.

Inputs

Over 5½ years, starting in 1972, expatriate technical help amounting to 15 man-years, with counterpart local staff, set up the installations and services. The latter included accounting, inventory control, supply, distribution, and after-sale service, as well as administration of the loans. More than 3 600 motors, rather than the 3 500 planned, were actually delivered.

Recruitment of local counterpart staff was slow, but eventually 35 mechanics and 10 storekeepers were trained.

From 800 to 1 000 replacement engines were needed each year.

Trained mechanics at the local centres, able to keep repair costs to the minimum, could extend the useful life of the engines, but there were delays in the delivery of replacement engines and spares. To keep a steady flow of these requires advance planning so that remote fishing centres have what they need when they want it. Not all the workshops were ideally sited, but they were built in an effort to get the program moving.

During the project three local management agencies were set up. These financially autonomous bodies were: a centre to help with the motorization of the canoes, a fishery development centre, and a limited liability company for artisanal fishery improvement. They received funds both from the lending agency and revenue from the sale of motors.

Outputs

Besides training personnel and operating the seven field stations, the project also gave short courses in preventive engine maintenance to a number of fishermen from the cooperatives. The administration that was set up gradually acquired the ability to manage the various services. However, the technical services were not as efficient as had been hoped. The project continued to need the services of expatriate staff and government help, and it became evident that the schedule of the project did not allow enough time for the services to be fully implanted.

The sales contract that the fishermen signed was flexible and could be a model for West African projects. Loans were secured by an insurance scheme against loss of the engines. These financial arrangements improved the stability of the artisanal sector of the industry.

The outboard motors were seen as a transitional stage to greater use of power in fishing, and a few of the larger canoes had diesel engines installed in them. Project staff conducted cost-benefit studies on different motors and different fishing methods, which provided a valuable guide for investment and credit policy.

Effects

The studies produced statistics that showed the improvement in catches. Some small fishing units that had been catching 10 to 12 tonnes a year harvested, with motorization, now as much as 35 tonnes. The project coincided with the introduction by FAO technical assistance personnel of a new, modified purse-seine system that used larger engines and three large canoes per fishing unit. These groups were able to catch up to 45 tonnes a year of pelagic fish; more than 100 such units are now in operation. Table I shows the effect of production on investment.

Table I RETURNS ON TOTAL INVESTMENT IN BOATS, MOTORS, AND FOR VARIOUS FISHING METHODS, 1978–80

	Hand lines	*Bottom gillnets*	*Surface encircling gillnets*	*Canoe purse-seines*
Number of fishing units	1 700	1 000	70	280
	%	%	%	%
Annual gross revenue as percentage of total investment	610	250	310	430
Owner's net revenue[1] as percentage of total investment	126	20	36	157
Owner's net revenue as percentage of annual amortized capital costs	500	59	144	559

[1]Owner's net revenue is calculated as gross return from fish sales less variable costs (fuel, maintenance, etc.) and less shares paid to fishermen.

There were difficulties with supply of spare parts. The success of motorized fishing created a far greater demand for engines and spares than had been foreseen, and there were gaps in supply from the far-distant overseas manufacturers. As a result engines, and therefore fishing units, were immobilized at critical times of the season. The fishermen found this unacceptable, and eventually the project administration sought supplies of another make of engine from a different country. A competitive manufacturer offered a gift of 1 000 engines and 'guaranteed' supplies of spares. In fact they provided only 600 engines. However, local dealers supported the change.

This produced a difficult situation, both for the donor agency itself and for the fishermen who had accepted the original engines, which could not now readily be kept in service. Maintenance service continued to be available from the trained management staff at the seven field stations, but the lead time required for ordering spares posed a problem, although local dealers carried stocks of the most needed parts – at considerably higher prices. The supply problem was not fully solved during the project.

The world-wide inflation in fuel prices occurred during the project, but fuel and gear supplied through the fishermen's cooperatives were tax-free. In 1983 the price of fuel for fishermen was only a third of that charged for road vehicles. Fishermen's costs thus were somewhat moderated, while the controlled retail price of fish showed only a mild increase.

To make motorization profitable for fishermen, there had to be a greater catch of fish to cover the increased capital and operating costs. This did in fact occur, but the market demand for fish was still not fully satisfied; thus the downward pressure on fish prices that would have resulted from a glut of supplies did not happen.

Impact

In the final year of the project, 1978, the consolidated accounts for the three organizations showed revenue of $2.06 million against expenses of $1.36 million. The revenues derived from sale of motors (60% of the total), spares (16%), fishing nets, seagoing weather-gear, and ice, plus repair services. Additional engines, replacement gear, supplies, spares, salaries, amortization, and other operational overheads accounted for the expenses.

The accounts therefore showed that the services were financially viable, but there remained a need for efficient control of operations. The commercial experience of maintaining an inventory of engines and supplies, by forecasting the needs six months in advance, was not easily acquired by the project administration. Local commercial dealers had the necessary expertise, and the needed spare parts were generally in stock there, at exhorbitant prices. This was a situation that did not inspire full confidence among the fishermen in the project as the sole source of supplies.

The project also intended that the cooperatives would later gradually develop the expertise needed for wholesale fish marketing. This would involve enlarged operations that would require even higher level of management with more commercial experience in handling this very perishable commodity, involving commercial operations for storage, marketing, and distribution of the catch.

Between 51% and 70% of canoes were motorized. Usage of motors varied by area of the country. The effect on production can be seen in Table II.

For fishing systems, as Table I indicates, net returns as a percentage of investment varies with different methods. For hook-and-line fishing, the

Table II MOTORIZATION OF CANOES AND ANNUAL PRODUCTION OF FISH, 1960–1980[1]

	1960	*1968*	*1971*	*1973*	*1974*	*1980*
Total canoes	3 900	5 140	5 293	5 930	6 442	8 556
Motorized canoes	702	2 646	2 408	3 561	4 187	5 561
% canoes motorized	18	51	49	60	65	65
Fish production (tonnes)	80 000	125 000	180 000	227 000	263 000	202 186

[1]Project began in 1972 and concluded in 1978.

investment in the engine yields an increase in the catch worth five times its cost. This results from being able to spend more time on the fishing grounds, and possibly having more manpower available for each canoe.

In the case of deepset gillnets or encircling gillnets, by contrast, fewer crew are needed and hence it is possible to have more gillnets in the boat, besides increasing productivity per worker. Often, however, this opportunity is not taken up because of the use of customary members of the crew.

The encircling purse-seine depends largely on the speed of the canoes in surrounding the school of fish. Accordingly, this new system is more effective when using a powerful outboard engine. This canoe purse-seine system accounted for the largest catches and the highest returns on investment. It is possible, however, that use of the system could over-exploit local stocks and some limitation of numbers in certain locations may become necessary.

The increase in production possible by using motors has led to motors becoming the standard method of propulsion and the main source of recurrent expenses, especially with the increase in the price of fuel. For small fishing units and canoes, the engine is the largest cost, whereas with larger canoes and nets its cost represents less than half the investment. However, even though it may be of less value than the larger encircling gillnets, the engine is critical, in that without it these operations cannot be carried on.

It is important that the studies done of the profitability and constraints of using engines with different fishing systems be well documented and explained to fishermen by extension agents. The impact of different systems on fishstocks also needs assessing and monitoring.

Working conditions have improved for motor-equipped fishermen.

Apart from the previously mentioned advantages of speed in getting to and leaving the fishing grounds, motorization has reduced the difficulties of getting to sea with paddle and sail. Fishermen, now commonly equipped with weather gear, undertake more frequent fishing trips and use larger canoes that can take more gear and land larger catches. Some have introduced large, insulated boxes in their larger canoes. However, there have been incidents when motors failed at sea and crews were not able to return safely. Instruction in preventive maintenance and repair has somewhat improved safety at sea, as has the mandatory carrying of buoyed safety lights and other safety gear (although this is not often used). A film on the use and care of engines is now available for fishermen's instruction. However, much remains to be done in providing continuous training in preventive engine maintenance.

Living conditions in the fishing villages have evidently improved. Many fishermen's families have better-built houses with permanent roofs, some with radios and even TV sets. Older fishermen express satisfaction that the younger members of the community are now more interested and take pride in fishing. This may result in greater social stability of fishing villages.

Whether there has been an actual direct increase in fishermen's incomes is not clear. Some fishermen assert that unexpectedly high costs of motors, fuel, and repairs, combined with revenue-losing delays while they wait for parts, have resulted in a net income that is not relatively higher, given inflation. Many suggest that middlemen – mechanics, fuel salesmen, processors, marketing people – have got more out of the increase in catch than have the producers. A set of simple guidelines in bookkeeping would help fishermen account for the costs and revenues in different fishing systems and resolve the question of costs versus benefits for different operating aspects.

Fishing cooperatives have spread the benefits of motorization. For instance, in one district 3 630 fishermen belong to cooperatives; between them they own 581 engines. In 1982 they spent $19 300 on motors and $22 500 on gear and landed more than $200 000 worth of fish.

Clearly the motorization of small fishing canoes has become an irreversible process. It would be useful to carry out a thorough socioeconomic study to determine how best to apply the benefits of this development to various communities and fisheries in any one country; a few successful operations are not enough from which to recommend

Fig 5 Fishing villages have many temporary seasonal dwellings

general application for use with all artisanal fishing systems.

Continuity regional centres. The project, then, resulted in better central and district fishery services and stimulated greater professional activities and earnings among fishing families. It provided increased production, employment, and food for the rural sector of the country.

It was thus a considerable achievement, and it provided experience to nationals of the country in the operation of new centres for fish storage, ice supplies, and redistribution. They thus gained managerial competence. The project helped develop cooperatives, through which came social and economic improvements in the fishing communities.

A follow-up project, lasting six years, was subsequently initiated. Though it was executed as a separate project, it concerned the disposal of the catch and thus might be considered as an integral part of the artisanal fisheries improvement program.

The follow-up project aimed to set up eight main landing areas, where the fish would be weighed, sold, and stored for onward distribution by refrigerated trucks. The fish would therefore be kept in better condition with less spoilage, and fish offal could be accumulated for appropriate use. There would be ample cold storage and ice-making facilities, sanitary conveniences, water supplies, and radiotelephone contacts with headquarters and inland marketing outlets. There would be mechanical conveyors for bulk washing of the fish; cleaning and storage of plastic fish boxes; and sections for packing, control, and shipment, with necessary service equipment. This contrasted with the traditional landing and sale of fish on beaches that had no sanitation and where the product was exposed to hot sun, rain, and unhygenic conditions. Fishermen who did continue to land fish on the beaches would be able to transfer it directly to the receiving stations.

The initial plan was that four of the centres would be set up between 1977 and 1980, the other four being established later. The project would cost $18 million, of which the donor agency was to contribute $12 million. There would be 22 man-years of expatriate management services and 20 training scholarships. The whole scheme would be managed from a central headquarters.

By 1983, three of the planned centres were operational. The delay and world-wide inflation greatly increased the costs. Wholesalers competed vigorously with the centres, not all of which received as much fish to handle as they might have done. However, their proportion of the available fish seems likely to rise as the motorization scheme takes hold and fishermen's cooperatives increasingly support the centres.

Evaluation and assessment

The project was able to set up the necessary back-up services for the motorization of canoes within the time scheduled. However, it seems clear that there was not enough time to allow the national management to become fully effective. This would have required longer service by the expatriate staff or more lead time recruiting and training local counterpart staff. The planning of the program might have included consultation with fishermen's community organizations, as during the project many details of it had to be modified to suit local conditions.

There need to be further studies on the limiting factors involved in motorization and on the benefits that the target population has derived. Given that motorization increased the landed value of catches by 20 to 50%, thus covering its own costs, it is necessary that for effective resource management government be enabled to determine the level of motorization desirable. This would be required for different fishing systems in various localities and would be assessable through continued monitoring and analysis of revenues and constraints. These data will also show any decline in catch per unit effort.

Difficulties encountered during the project included the slow recruitment of suitable staff and the slow pace at which they were taught to operate the rural centres. Training of support staff was incomplete, a problem still not solved by 1983. Collaboration between the government department responsible for fisheries and other departments was poor, a factor that in particular hindered the setting up of the motor service stations. Radiotelephone links with headquarters did not function properly so that the communication of urgent needs was hindered.

The donor agency could have been more flexible by allowing expatriate staff to serve longer than the agreed periods. The agency adhered strictly to the agreement, which the host government did not. Thus agency representatives in the country might have introduced measures to ensure local preconditions were met. Possibly the agency expected a better level of local management and field personnel than it was possible to recruit. Perhaps also neither agency nor government at the planning stage realized the degree of private competition the project would meet; had they done so they could have enlisted the support of local representatives of the motor manufacturers. None the less, the private sector did have positive contributions to make to the development of artisanal fisheries.

Project management seems to have been satisfactory in the technical field operations. The management, however, compromised on the siting of the service stations, which were not all located conveniently for the use of fishermen. The system for ordering spare parts could not cope with the difficulties in local and overseas communications.

The host administration's recruitment of staff and acquisition of field sites was in general slow. The government also was in financial difficulties and found it difficult to produce the necessary money to finance its obligations. Despite the high priority it evidently placed on development of artisanal fisheries, it was slow to respond when critical issues arose. Interdepartmental cooperation and liaison with fishermen might have improved had there been an advisory committee. The government did support its stated policy with tax incentives, but its field services were not adequate to provide effective support continually. Government needed to follow up the specific results of the project with effective management strategies in different localities of the coast requiring the limitation of numbers of fishing systems. Equally important was the involvement of private commercial traders in the changed operating systems being introduced.

Continuity: The motorization project, the cooperatives, and the maintenance centres have continued to operate with varying degrees of efficiency. Some 8 000 fishermen belong to 80 cooperatives, and the volume of business in supplies, gear, engines, and spares is sufficient to sustain viable operations, although greater management skills are needed. The marketing centres that were being established would need even more experienced management to operate as self-sustaining cooperatives in the face of private competition; these centres might preferably have been built at the same locations as the motor service centres, thus forming integrated service centres for artisanal fisheries.

Management considerations: The project demonstrated for West Africa that putting outboard engines on canoes is a practical and profitable means of increasing production in artisanal fisheries. This can, of course, give rise to resource conservation problems subsequently, although motors enable the canoes to fish wider areas. None the less, reliable data on size of catch and costs are needed to assess trends in fish availability and reach sound decisions on the amount of investment in such projects that is prudent for optimum harvesting.

Motorization can only succeed as a long-term project with proper back-up service for the motors, instruction in maintenance, and availability of spares. The increased production can lead to localized oversupply of fish, which can result in downward pressure on prices from middlemen, unless adequate cold storage, distribution, and market arrangements are in place.

Staff: Top priority must be given to recruitment and further training of national management staff with experience in commercial operations of this type. Efficient service mechanics also need continually to be trained. There should be time for the local staff to reach the necessary level of expertise before the expatriate staff leaves. Adequate instruction to the operators of the motors in preventive maintenance can reduce the foreign exchange costs of buying spare parts.

Planning: During the planning of such a project all the operations must be considered together. This includes the necessary extension services – training in the use of engines and collection of data on costs and returns – as well as the proposed facilities. Preparation must include the well-planned involvement of beneficiaries and competitors, the phased execution of each stage of the project, and long-term, consistent government support. Scheduling must be realistic. It should allow for pre-project recruitment and training of extension personnel and refresher courses. No new service centre should be launched until competent management is ready. Problems that arise should be solved so that lessons learnt can be applied to new phases of the project.

Records: The extension service could also provide artisanal fishermen with a simple bookkeeping system, with records of costs of engines, maintenance, fuel, gear, and boats and of revenues from the catch. This will enable them to understand which items need to be watched so as to ensure profitable fishing. Different localities, different fishing methods, and different seasons will all produce varied results of profit-and-loss.

Spares: Timeliness in the ordering of spare parts is critically important. Personnel need to know about parts numbers, interchangeability of parts if more than one model of engine is in use, and design changes in engines, otherwise there will be confusion in ordering and waste. Well-trained mechanics can help by advising on the parts needed and sometimes by adapting parts to keep an engine running. In the project under consideration many engines were abandoned for want of a small part, with the repeated result that fishermen lost confidence in their service centre and went to dealers who sold another make of engine.

Diagnoses: A great deal of attention has to be paid to the problem of keeping engines running. Why do they break down, and what can be done by way of regular maintenance to prevent this? Obviously, the new parts most frequently ordered offer a clue to any pattern of misuse; based on this information, some additional retraining of mechanics and owners can help ease the breakdown problem. For instance, it is necessary to reduce corrosion by seawater of aluminium alloy components by regular washing and greasing, but often there is insufficient domestic water in the fishing villages; accordingly more communal washing and test tanks are needed. This simple, nonmechanical factor would help extend the life of motors, which at present is rarely longer than two years.

Services: Fisheries service stations should be sited according to clear criteria – on the landing beach, with water, electricity, and telephone communication. To place them elsewhere causes hardship to the fishermen and defeats their purpose. All the facilities for a general fishery centre should be located together, for greatest effect and to avoid duplication. Each coastal community is unique, as a result of tribal customs, fishing methods, seasons, geography, climate, and the impact of industrial fishing, foreign fleets, nearby ports, and other neighbouring factors. Thus each fishing centre will have its own special character, but each should have the same basics – efficient sales and service for engines and gear and market facilities for fish. Each centre also should have its own local committee to review operations, discuss and solve problems, and generally support the integrated services.

Institutions: It may be that a mix of public and private enterprise can best address the needs of fishing communities. For instance, fish trading and credit by a dealer network are pervasive in West Africa. It may be more effective for governments to provide the necessary improved installations and for dealers – and eventually cooperatives – to rent them and conduct marketing operations. Universities and local high schools could help conduct the studies necessary to gather baseline data on catch, operations, and socioeconomic conditions from which to monitor the progress of the project. These data, with other data compiled by the extension service of the fisheries service, can provide the basis for formulating policies for the practical management of the resource. Where there are local committees, the data can be explained to them, thus gaining support for any management measures necessary.

Lessons learnt

Summing up, then, there are lessons to be learnt from this project and applied to similar schemes to motorize traditional fishing fleets. The lessons can be applied in the early stage of planning and negotiation, at the stage immediately before operations begin, during implementation, and after the project is ended.

At planning stage:

- Ensure a review of results and difficulties in similar projects in nearby countries by technical staff to obtain practical operating guidelines.
- Check that the overall project plan is realistic, that there is staff able to implement it, and that the schedule is feasible, given local conditions and current infrastructure.

- Confirm that the host government is clearly determined and able to meet all the commitments it has undertaken.
- Ensure fishermen and local authorities are involved in planning through local and central advisory committees.
- Ensure collaboration with local representatives of engine and gear manufacturers.
- Ensure collaboration of all branches of the fisheries service and other government departments.
- Identify the project manager and the senior national staff who will work in the project.

At immediate preproject stage:

- Establish clear, inflexible criteria for the location of field operational centres.
- Establish standards for the recruitment of local management and ensure their training is adequate.
- Ensure the expatriate project manager has experience relevant to the area and is familiar with difficulties that similar projects in nearby countries encountered.
- Ensure that the expatriate and national project managers share a rational process of decision-making.
- Organize a realistic schedule for construction and installation of equipment in view of local capabilities.
- Establish a central advisory and coordinating committee of all significant branches of government and local services to review regularly project activities and solve administrative difficulties. District advisory committees may also be established.
- Set up an independent monitoring and evaluation system through a competent and interested local body (such as a university or research institute) to assess data and transmit results to the extension service.

During implementation:

- Organize the continuous training of fishermen by project and extension staff. (Extension leaflets and posters that may be needed are best prepared locally.)
- The advisory committee and monitoring team should present regular half-yearly reviews.

- A mid-project review with national policymakers, in addition to the above, could help stabilize project activities.
- Management should preferably overcome pressing problems, such as parts replacements, communications, or specific local difficulties, before establishing new, more distant centres.
- Management should continue collaboration with local representatives of engine manufacturers, as their activities can help or hinder the project.
- Data on the performance of engines and fishing units should be collated annually for the information of management, government, and the public.

At project termination:

- A published analysis, assessment, and evaluation of operations would be valuable for other, similar undertakings. It should give a breakdown of the economic, social, and nutritional benefits that the program spread directly and indirectly among the various beneficiaries. The report should be prepared mainly by the national project management, but with independent inputs from the monitoring and evaluation unit. Such a report could be part of the national and regional experience, could guide bilateral and multilateral programs, and could communicate the lessons of the program to all those concerned with fisheries, world-wide.

Selected references

Allsopp, W. H. L. 1966 (TOGO) Développement et organisation de l'industrie des pêches. Rome, Italy, FAO, FAO/UNDP (TA) report 2184/F. 19 p.

1970. (ECUATORIAL GUINEA) La situación y posible desarrollo de la pesca. Rome, Italy, FAO, FAO/UNDP (TA) report 2832/S. 30 p.

Aubray, R. 1970. La pêche Maritime au Maroc. Mimeograph. Rome, Italy, FAO. 105 p.

1971a. The fishery resources of West Africa. Compilation. Mimeograph. Rome, Italy, FAO. 440 p.

1971b. The fishing industry of Morocco, summary of a study. Mimeograph. Rome, Italy, FAO. 23 p.

1972. Mauretanian fisheries. Mimeograph. Rome, Italy, FAO. 41 p.

1975. The fishery of The Gambia. Mimeograph. Rome, Italy, FAO. 24 p.

Bayagbona, E. 1975. Fisheries in Nigeria. Nigeria Trade Journal, 21(1), 6–12.

BCEAO (Banque central des Etats de l'Afrique de l'Ouest). 1972. La pêche au Senegal. Econ. Ouest Afr. 193, 1–25.

Ben Yami, M. 1971. Ivory coast. (Experimental purse seining with lights and a fishing survey in the open coast.) A report prepared for the survey and development of pelagic fish resources (*Sardinella*) project. Mimeograph. Rome, Italy, FAO, FI:SF/IVC/6/1. 26 p.

Bowman, H. W. 1955. The encyclopedia of outboard motorboating. New York, USA, Carnese Co. 424 p.

Collart, A. 1965. (DAHOMEY) Organisation et développement des pêches. Rome, Italy, FAO, PEAT/FAO rapport 1945/F. 85 p.

1967. (DAHOMEY) Organisation et développement des pêches. Rome, Italy, FAO, FAO/UNDP (TA) Report 2302/F. 20 p.

Corten, A. 1971. Some data on landings of pelagic fish by artisanal fishery near Freetown during the period March – December 1970. Mimeograph. Rome, Italy, FAO, Report on survey of the development of pelagic fishery resources, UNDP (SF)/FAO, (71/7). 7 p.

Crutchfield, J. A., Lawson, R. 1974. West African marine fisheries: alternatives for management. Washington, D.C., USA, RFF, RFF/PISFA Paper 3. 64 p.

Di Palma, S. 1970. The marine fisheries of Morocco (including developments in 1969). Washington, D.C., USA, Department of Commerce, Foreign Fisheries Leaflet SFWS 40. 15 p.

Dogny, J. A. 1971 (DAHOMEY) L'organisation et la développement des pêches. Rome, Italy, FAO, FAO/UNDP (TA) Report 2949/F. 32 p.

Ezenekwe, A. O. 1974. Nigeria develops fishing to meet food demand. Fishing News International, 13(9), 42–49.

FAO (Food and Agriculture Organization). 1966. Takoradi, Ghana, in Landing and marketing facilities at selected sea fishing ports. Rome, Italy, FAO, in FAO Fisheries Report R36/E, 289–298.

1967. Report of the FAO Technical Conference on the fisheries of West African countries, Dakar, Senegal, 31 July – 4 August. Rome, Italy, FAO, FAO Fisheries Report R50/E. 23 p.

1971. Report of the consultation on the conservation of fishery resources and the control of fishing in Africa, Casablanca, Morocco, 20–26 May, vol. 1. Rome, Italy, FAO, FAO Fisheries Report R101/E. 28 p.

1974a. Eastern Central Atlantic (Area 34 – CECAF): availability of catch statistics. Rome, Italy, FAO, FAO Fisheries Circular 464. 40 p.

1974b. Eastern Central Atlantic (Area 34 – CECAF): classification of aquatic animals and plants for statistical purposes. Rome, Italy, FAO, FAO Fisheries Circular 461 Rev. 1. 35 p.

1975. Task force report on the marine fisheries development prospects of Morocco. Rome, Italy, FAO, FAO Investment Centre Joint Working Group. 13 p.

Undated. Working seminar on coastal pelagic resources of West Africa, Tema, Ghana, 1–5 November 1971. Rome, Italy, FAO. 64 p.

Fiastri, G. 1971. Small boats and their motorization for artisanal fisheries. Rapport Information du Projet du Développement de la Pêche Cotière. Abidjan, Ivory Coast, FAO/PNUD/FS 3/71. 12 p.

Ivory Coast: Ministère de la Production Animale. 1969. Plan schématique d'action pour un programme de développement de la pêche artisanale. Rapport Information du Projet du Développement de la Pêche Cotière. Abidjan, Ivory Coast, FAO/PNUD 6/60. Unpag.

Lawson, R. M., Kwei, E. A. 1974. African entrepreneurship and economic growth: a case study of the fishing industry in Ghana. Accra, Ghana, Ghana Universities Press. 262 p.

Lefebvre, R. 1975. Handbook of artisanal boatbuilding. Rome, Italy, FAO, CIFA/2. 131 p.

Mansvelt-beck, J., Sterkenburg, J. J. 1976. The fishing industry in Cape Coast, Ghana. A socio-economic analysis of fishing and fish marketing in a medium-size African town. Utrecht, Netherlands, Geog. Inst. Rijksuniversitett. 61 p.

Nguyen van Chi, B. R. 1981. L'essor de l'économie de pêche artisanale et ses conséquences sur le littoral sénegalais. Cahiers d'études africaines. Seg Jnl Source, 20(3), 255–304.

Pease, N. L. 1972. Fisheries of Mauritania, 1971. US Department of Commerce Foreign Fisheries Leaflet 73-7. 9 p.

1973. Fisheries of the Ivory Coast, 1972. US Department of Commerce Foreign Fisheries Leaflet 73-19. 15 p.

1974a. Fisheries of the Cameroons, 1973. US NMFS/NDAA Foreign Fisheries Leaflet 74-12. 12 p.

1974b. Fisheries of Ghana, 1972. US NMFS/NDAA Foreign Fisheries Leaflet 74-2. 12 p.

Stequert, B. *et al.* 1979. La pêche artisanale maritime au Sénegal: étude des resultants de la pêche en 1976 et 1977, aspects biologiques et économiques. Orstom, Dakar, Senegal, Inst. sénegalaise de Recherches Agric Dakar, vol. 73. 48 p.

Thomas, A. J. 1965. Mechanization of small fishing craft under revolving fund arrangements in developing countries. Rome, Italy, FAO, FAO Fisheries Report R24/E. (Also available in French and Spanish.) 28 p.

1969. (THE GAMBIA) Further development of fisheries in The Gambia. Rome, Italy, FAO, FAO/UNDP (TA) Report 2752/E. 19 p.

Traung, J-O. 1963. Mechanization of fishing craft. Rome, Italy, FAO, FAO Fisheries Paper 22/E. 12 p.

4
Institutional training for small craft construction in Central America

Background

Next let us look at a project in a Latin American country to improve the standard of fishing boat design and building. The country has a coastline on the Atlantic and Pacific oceans and its population is sizeable for that region. More than 30 000 fishermen, mostly part-time, form the artisanal sector, operating in coastal and estuarine waters at many points. Accurate information on their production is not available. There is also an industrial fishery for shrimp and pelagic fish. Total fish production was estimated before the present project began at more than 60 000 tonnes a year, and per-capita annual consumption of fish in the country at 3 kg to 4 kg.

The artisanal fishing boats were mostly small canoes – less than 10 metres – most of which used paddles and sails, although about 30% had outboard motors. Gear is rudimentary – miniature longlines, traps, and gillnets for catching crabs and estuarine or coastal fish. There is little supporting infrastructure for the artisanal fisheries. Development possibilities depend largely on better management of the resource, with improved craft so that artisanal fishermen can more efficiently harvest the abundant stocks.

Between 1965 and 1975 the country's development programs included projects to organize both industrial and artisanal fisheries. The government sought aid, and projects were undertaken with bilateral donor support as well as with help from multilateral donors and lenders. Regional and national banks invested in the industrial sector, and this stimulated shrimp trawling by private enterprise on both coasts.

The artisanal fishery consisted of various fishing systems in coastal bays, estuaries, and rivers. The types of seafood caught included mackerel, croakers, shark, shrimp, crabs, and shellfish. Most of the fishing took place at night so the catch could be sold at the market early in the morning. Typically there was inadequate preservation of the catch, although some ice was used, and in the more distant communities some fish was dry-salted. Catches varied according to lunar cycles and spawning seasons; the market was usually undersupplied.

Artisanal fishermen suffered some competition from industrial operations in coastal waters. Shrimp trawlers at sea caught fish as a bycatch and sold it to artisanal vendors, thus depressing somewhat the market.

Objectives

During 1972, an exploratory mission from a bilateral donor agency assessed the condition of boats, gear, fishing zones, commercially exploited species, and markets. Subsequently the project described in this chapter was initiated by the donor agency to provide instruction in fishing and boatbuilding at an already established vocational training centre on the Pacific coast. The program had five facets:

- To set up an administration to run the project.
- To train local teachers and thereby improve instruction at the centre.
- To introduce new courses for small craft design and construction in wood and steel.
- To establish workshops with the tools and facilities needed for on-the-job training in boat construction and maintenance.
- To help design courses and prepare instructional materials in fishing systems.

It was also planned to introduce a practical seagoing course to train fishermen, but the donor agency subsequently provided this part of the program in a separate project in fisheries development.

The project also included choosing a boat design to replace the dugout canoes then in use. The favoured design was a 10- to 12 metre boat, seaworthy in rough swells, with an effective working platform, that could use various methods of fishing, such as longlining, gillnets, seines, traps, trolling, and pair trawling. It should have an insulated icebox hold of one to two tonnes' capacity and, preferably, an inboard engine that could be sure of local maintenance through reliable agents and supplies of spare parts. One of the standard designs of FAO met the specifications; it would cost, with engine, not more than $6 500.

The assessment showed that the artisanal fisheries were at a low level of organization, and materials and services were limited. There was little modern equipment available and the communities showed the kind of

Fig 6 New design craft built at a trade school

low technology and self-sufficiency seen in isolated pioneer communities around the world. Artisanal boatbuilders within these fishing villages were working by tradition and without plans. This construction system is no longer cost-effective for the production of fishing craft.

There was a shortage of good quality wood, on which the costs would depend, and in remote areas there were no wood preservatives and antifouling paints available. These were essential for use in boat hulls, so as to deter wood-boring molluscs. Available wood was also in demand for furniture, housing, and the export trade, so poorer quality wood was often used for boats.

In most villages no electrical tools were used, and each boat was handbuilt from the keel up. Productivity was low, the boats were expensive, and there were few buyers. With little turnover there was little skilled labour, which resulted in few permanent employees, not much transfer of skills, and low quality of workmanship. Thus, the need for seaworthy boats was largely unfulfilled. Further, there was no insurance readily available for larger or better-built artisanal craft and there was difficulty in getting bank loans to finance them. It was proposed that the introduction of boats of better design would be carried out by builders trained at the school. There, a number of features, including plan reading and pattern making, would be introduced.

The project therefore would include a program of training for instructors. Wooden vessels were to be constructed at the centre, for demonstration, not for sale. In view of the materials problem, steel was a realistic alternative, and the program would include the construction of steel vessels. The artisans trained in this program would form a nucleus of private boat builders. Training would include welding and machining, as well as carpentry. Only the last-named had previously been taught at the centre.

Inputs

The expatriate staff and the equipment that the donor agency would provide were clearly set out in the agreement between the agency and government.

The donor agency would contribute training equipment for construction, navigation, and fishing methods, an equipped training vessel, scholarships and language training, technical experts, and administrative services. This, with miscellaneous expenses and an allowance for contingencies, would cost about $2.4 million.

The host government would contribute the cost of the boatyard, some equipment and furnishings for the demonstration vessel, internal transport and installation of equipment, wharfage, and office furniture and supplies. Also, for two years it would pay the costs of operating the training vessel and provide necessary administrative services and the staff and supplies for boatyard construction. With an allowance for contingencies, this would total the equivalent of $0.63 million.

Outputs

The project was planned to continue for three years, and was actually continued for two further years. At the end of that time not everything planned had been achieved – the national staff had not all been appointed, all buildings had not been erected, and the training vessel had not been delivered, as the host government had not accepted a modified vessel on the grounds of unsuitability. After the five years, three wooden boats of less than 10 metres had been built, and a 12 metre boat was still under construction. Training in boatbuilding with sheet metal had begun and the necessary equipment for this was in place. The industrial boatbuilding yards were collaborating in the project and the project management was being improved.

At this point it seemed likely that the skills that were being developed at the centre would spread along the coast. Accordingly there was a

Fig 7 Frame of a fishing boat

proposal to extend the project, possibly financing it from a levy on the established boatbuilding yards.

Effects

Three staff instructors at the centre received professional training, and more than 20 students took part in building the three small wooden craft using the new electrical and other equipment supplied. However, the management of the centre did not regard the construction of fishing vessels as a major activity in comparison to production of other types of coastal vessels. Furthermore, the centre was controlled by a government department whose headquarters was at a considerable distance from project operations. Although the department showed enthusiasm initially, its level of support seemed to wane, and it was doubtful the program would survive after the withdrawal of expatriate staff. It is possible that the centre management used the project as a means of getting more funds and staff overall, but the implementation of project objectives did not directly benefit from this.

The training program was considered effective and some of the trainees showed considerable promise. However, for the program to have a lasting effect the training should have been more intensive, with a rotation of instructors to vary the experience the trainees underwent.

The low status of coastal fishing operations meant there was no great public demand for the new services that the program made available. The absence of the planned training vessel caused a lack of publicity for the project. Thus the effect on local boatbuilding or the fishing community was small.

Poor administration and a number of incidents (a floating dock of ferroconcrete, not part of this project, was lost at sea, for instance) were bad for the morale of project staff. The expatriate staff did not have time to adjust to conditions and pass on their skills. It took time to organize a working team that would, within a somewhat rigid administrative structure, conduct an autonomous, well-managed activity. There were long periods for preparation and organization and for the selection and delivery of equipment. This cut into the time available for the training process.

Because the centre is remote from the host agency's headquarters, there were shipment delays. It is clear, in retrospect, that the original plan of a cut-and-dried 36-month operation (later stretched to five years) was impractical. Nor was it realistic to impose a boat design that had been satisfactory in other countries but had no local testing; indeed, none of this model was subsequently built by commercial boatbuilders.

The donor agency was therefore faced with the dilemma of whether to abandon the cash investment already made or whether to put in more money to bring the project to the point where it could usefully be handed over to national management. One suggestion was to change the project and add a new management system with local counterparts, a revolving fund for operations, continued support for existing activities, and improvements in the physical installations. More time would be allowed for procurement, and follow-up activities would include teaching materials.

Impact

Because there was no prior liaison or later arrangements for the placing of trainees, managers of small boatyards felt the scheme was of little relevance to them. They agreed they needed trained staff, but said there were not enough orders for new craft to justify skilled tradesmen. A greater demand for fishing vessels would have to be stimulated by a loan scheme and insurance. They felt the training, though desirable in itself, was not enough in the absence of an integrated development scheme; improved fish harvests, without vastly improved preservation and

Fig 8 Fleets of steel trawlers also require shipyard maintenance

marketing facilities, would be pointless. Thus costlier craft or an increase in catch created a chain of problems in which small fishermen are trapped.

The government's fisheries services seemed to have little liaison with this project. Separate departments of government were responsible for technical advice or extension, cooperative organization, social services in fishing communities, and fisheries regulations or controls. These departments should have had input into the design and trial use of the fishing craft built by the project, but they didn't. A coordinating committee had been planned as part of the project, but it did not function.

There seemed to be a long chain of command in departmental services. The decision-makers at the head of this chain did not fully understand the field operating problems put to them and were not able to resolve them. Field activities suffered from a lack of local coordination and delegation of authority to resolve problems. This discouraged both expatriate and local workers on the project.

Evaluation and assessment

This was one of several development projects carried out by the donor agency, and it represented only 4% of the agency's support given to this country. The failure of the donor to provide an acceptable training vessel caused some embarrassment and awkward relations between the expatriate and national staff of the project. Thereafter the donor agency seemed anxious to placate the host government and did not succeed in persuading it to meet its commitments to this project. Donor agency staff seemed to feel that the agency hesitated to upset the harmony of its overall program for one small project. Not only was the agency indulgent to the government's lapses, it was itself slow to react to the difficulties that became evident. There was, for this project, no expatriate manager, and the donor agency took the view that the project should be self-managed with as little interference as possible. There were many changes of national supervisory staff.

The national administration for this project was in the hands of a government department that was responsible for many development activities in widely separated locations. The management level was administratively and physically distant from the operational centre and did not show much effectiveness in dealing with difficulties encountered. Eventually the donor agency brought in a respected consultant, who

Fig 9 Much fish is caught by gillnets

reorganized the project, modifying the plan of activities to meet the current operating conditions with a practical action schedule.

The training centre is geared to instruction, but funds to continue the training were not available to the host department. However, once the money problem is solved, the centre will continue to give some courses in fishing and boatbuilding. None the less, its relevance to the fishing industry is in doubt pending an integrated development of the fisheries. Furthermore, unless the trainees are used in industry or helped to set up their own businesses through loans, the training impact will be lost.

A plan for the integrated development of the coastal area has been announced but not yet implemented. Until this happens it is unlikely that the introduction of a few trained fishermen and boatbuilders and a few new fishing boats will have much effect on the artisanal fisheries. Indeed, the centre's efforts under such conditions might result in discouragement of the trainees.

The craft that have been built have had no critical test of their fishing performance, economic returns, and seaworthiness in the coastal fisheries when in use by fishermen beneficiaries. Their construction and operational costs or production efficiency, as well as their social acceptance by the artisanal fishing community, have not been assessed. Where they will fish, in relation to the dugout canoe and shrimp trawling areas of operation, is unclear, since management or regulatory zones are not demarcated. Further, there seems to be no directly stated relation between the type of boat selected and the specific community usage (of gear, storage, duration of trip) to justify the cost compared with an outboard-powered canoe. The craft and the fishing systems it will use must therefore be tested by the fishermen who will use them before such a prototype is generally promoted as an efficient, acceptable fish production unit.

Continuity

The future of the activities discussed in this chapter is unclear. The centre hopes for further financing from the donor agency, but this clearly needs to be done as part of an overall program of fisheries development. International banks have been invited to consider funding the development of coastal fisheries, with perhaps several extension centres, expatriate staff help, and supply and service of gear and equipment. The country's government is examining ways of setting up a program to include biological research of the resource, economic studies, and financing of better processing and marketing facilities, as well as controlled zones for fishing. The centre could conceivably contribute to such an integrated program through training fishermen and introducing craft of better design.

Lessons learnt

The project offers us four major lessons:

- It is essential to have the administration of a project fully organized before starting it. Where the chain of command does not allow sufficient autonomy to local management, particularly in climatically difficult and geographically isolated areas, procedural problems tend to be more difficult than the technical work to be done. Host country commitments must be met.
- A project advisory committee should be organized to coordinate activities and resolve operational difficulties during implementation. It may be of critical importance in keeping the project moving.
- The schedule for procuring equipment, recruiting and training staff, and carrying out the various parts of the program should be set to take account of the locality of the project, communications, and other local circumstances and traditions.
- Where conditions change radically during the life of the project, it may be preferable to terminate the project and renegotiate another with a better likelihood of achieving its objectives. To prolong a project in which nothing is actually happening is counterproductive for donor and host, wasteful, discouraging for national staff, and prejudicial to future projects.

5
Tuna vessels, shore facilities and training in Latin America

Background

This chapter reviews a project to build and put into operation a number of boats much larger than are usually used by artisanal fishermen. The project was sited in a small Latin American country with a population of a little more than five million and a per-capita income of less than $200 a year, at the time the project was conceived. The economy was agriculture-based, and 80% of exports were agricultural products.

The artisanal fishery of the country consisted of canoes and small, inshore fishing boats; some 300 vessels of six to 32 metres were motorized. Total production in 1967 was about 53 000 tonnes, and fish products accounted for $6.3 million, or 3.4% of foreign exchange earnings. Fisheries development was constrained by inadequate port services, insufficient storage and distribution facilities, and, because of high transport costs, high prices in the main inland cities with a consequent lack of demand. There were an established shrimp and tuna fishery and a small fishery for spiny lobster and demersal reef species.

In 1965 about 17 000 people were employed in fishing, of whom 1 700 were in industrial fishing and a like number in processing the industrial catch. The country in 1968 took some measures to stimulate the private sector, offering a share of the abundant fish resources off its coast. These included skipjack, yellowfin tuna, and other pelagic fish such as mackerel and sardine.

The administration of the fisheries was divided between several authorities. The fisheries department was part of the ministry of industry and commerce; fisheries planning came under the national planning board; regulation and patrol came under the ministry of defence; and the ministry of health dealt with quality control. The fisheries department was newly organized and had a small corps of professionals and support staff. There was no extension service, although the need was recognized.

A national fisheries research institute was setting up a training centre for fisheries personnel.

The offshore fisheries were being exploited mostly by vessels of other nations, including about 60 vessels based in the country that for the most part were operated by large international companies. The government saw the attractiveness of the international market for tuna and adopted a policy of encouraging persons and groups within the country to own fishing vessels. If national crews could be trained to operate modern boats, and canneries were to export the catch, this would produce favourable results on the economy. Accordingly, legislation was enacted in 1973 under which a majority share in locally based boats had to be held by nationals.

Fig 10 Many countries have now established fishermen's training institutes

Objectives

In 1968 an agreement was signed between the country and an international lending agency for a project that would provide the construction, equipment, and operation of 12 tuna purse seiners at a cost of $5.3 million; a training program to cost $400 000; and studies for the development of new harbour facilities at a cost of $200 000. The training program would use expatriate staff to teach nationals new methods of fish capture, using a second-hand vessel purchased locally for the purpose. A team of foreign consultants would conduct the harbour studies. The schedule for the project was:

Year one: Designing the 12 purse seiners; recruitment and preparation of the expatriate training staff.

Year two: Start construction of the vessels; start training program and harbour studies.

Year three: First project vessel to be operating and under study to see what modifications would be needed in successive vessels.

Year four: Construction of five new purse seiners.

Year five: Construction of four new purse seiners.

Year six: Construction of remaining vessels.

Inputs

The proposal was that the lending agency should make the funds available to a national finance corporation. After the agreement was signed in 1968 there was a period of further protracted negotiations before implementation. Eventually the first disbursement was made only in 1973, five years after signature. Local investors then proved somewhat indifferent toward the loan funds available for purchasing vessels. The fund intended for the training program was transformed as the contribution to construct a permanent fishery school, eventually finished in 1976. The harbour studies dragged on. After several missions and revisions, the experts eventually concluded it would not be feasible to construct a new, central fishing harbour for the fleet, complete with all necessary facilities.

Outputs

Only four of the 12 purse seiners envisaged were actually constructed and put into operation. Although at the appraisal of the projects, these boats had been expected to cost $350 000 each, they actually did cost between $600 000 and $800 000 each. The lending agency cut its appropriation to $3.8 million. Meantime, the new legislation for fishing operations within the EEZ encouraged the entry to the fleet of 56 new purse seiners, none of which was financed by the project. Evidently there had been little participation in the design and implementation of the project by the private sector, some members of which may even have been unaware of its aims.

The government's other programs to boost investment in the industry thus led to so much growth that the capacity of the fleet was greater than the available fish stock could sustain.

The training program had been intended as an on-the-job scheme aboard vessels with expatriate captains. In its new form, as a training institution, it received substantial bilateral aid from sources not connected with the international agency. It became well established, serving

Fig 11 A mobile audio-visual unit for a coastal fishermen's training institute

both the tuna fleet and the artisanal fishery; by 1982 nearly 600 students had received training and its graduates were to be found on most of the larger tuna vessels, though it also catered to the artisanal sector.

While the harbour studies were dragging on, two local anchorages gradually increased their importance through the steady commercial growth of the industrial sector. These anchorages became also used by the artisanal fishing craft. They do not, however, have dockside service or storage facilities. Various aid agencies are considering projects to build two small fishing ports and four landing facilities for artisanal fishermen. At one point the harbour study recommended developing one of these small fishing ports rather than constructing a new industrial harbour, but by the end of the implementation period the economic feasibility was in doubt.

Effects

The new laws requiring that nationals of the country hold a majority share of the ownership of each craft fishing within the EEZ came into effect during the project, and this situation was not foreseen when the project was being designed. The size of the investment required per vessel built by the project made it impossible for any one local entrepreneur to invest in a tuna fishing vessel; this led to joint ownership of such larger seiners. The four vessels built were relatively small and not very efficient or suitable for local operating conditions, with a limited operating range. Hence it was not practical to operate them without a mother ship, and it was obvious that larger vessels were needed. This further reinforced the difficulties of individual investment.

The seiners were individually built by external boatyards through international competitive bidding, and it would clearly have been better to have had them built at the customary tuna seiner shipyards, where the prospective owners could have first inspected similar craft and where the builder would have given the normal after-sales shipboard service, training, and maintenance.

The recommendations from the harbour study took 12 years to arrive after the signing of the project agreement. During that time, costs inflated, and a number of chances of getting bilateral aid for one of the two fishing ports were lost.

The international lending agency had little previous experience with this type of project and clearly gained experience in this instance. This

Fig 12 Purse seiners and motorized small-craft need safe harbours

led to better arrangements in its subsequent fishing projects. The national loan agency became disillusioned and may or may not continue to support fisheries investment. The government continues to promote fisheries and to solicit aid to do so. The fisheries administration has established a research and management institute, which assessed available fisheries resources and introduced management measures for the industrial fleet.

Impact

The impact on the fishing community of the four vessels actually constructed was minimal, especially as these seiners were in fact acquired by multinational corporations, as no local people had the resources and expertise to buy and operate them. Government services may, during the life of the project, have improved. The well-equipped training school will no doubt contribute substantially to the fisheries, especially as it is proposed to offer an extension program to fishing villages. It is even training mechanics to service coastal shrimp farming.

Although the project thus did not have much impact on either the country or the region, development efforts in neighbouring countries have motivated the government to seek further aid.

Private tuna and shrimp fishing enterprise has been considerably enlarged, and annual total production of fish rose during the period of the project from less than 60 000 tonnes to more than 500 000 tons; investment in the private fisheries sector has increased to nearly $1 billion. Little of this development can be directly attributed to the project; none the less, the increase shows that the decision 15 years ago to invest in fisheries development was valid; it can be said that development proceeded despite the difficulties that so hampered the project.

Evaluation and assessment

The final review by the international agency concluded that such programs should involve the private fishing sector in policy decisions that comprehensively affect all the fishing industry. An advisory committee to review progress could be a suitable forum for such consultations. In the present case, the government might consider appointing private sector members to the board of the training institute, inasmuch as that sector benefits from the training programs.

This was among the first fisheries loans provided by the international agency, which thus had much to learn about identifying needs, formulating plans, preparing projects, and making such loans. The agency learned that realistic targets should be set at appraisal; those set were too ambitious and the schedule was far too optimistic, bearing in mind what was normal for this country. The schedule should also make allowances for unforeseen delays. The contracts with consultants did not produce the hoped-for results, and such contracts obviously need more careful scrutiny than they received in this project.

After the loan had been approved, there was a rapid increase in foreign interest and investment in the tuna fishery. The private sector was, in fact, efficient and capable, something the appraisal mission seems not to have realized. Meanwhile the lending agency was dealing with the government and one national agency not directly responsible for all fisheries, and running into many difficulties. These included inflation, decisions on suitable types of vessels, the siting of the fishery school, and the extension of the EEZ. It would have been helpful to get tax concessions generally established on the importation of some items required for the fishery program.

These problems added to the delays and thus reduced the effectiveness of the project. The lending agency was generally indecisive

Fig 13 Launching in the surf is often hazardous

Fig 14 Without anchorage boats are stacked on beaches

and there were times when a more decisive action could have moved the project forward more effectively. The government felt that it did not get the prompt and constructive help and advice from the agency that it urgently needed; accordingly it turned to bilateral aid agencies to make progress. Some of the expatriate staff performed inadequately.

The fact that a number of arms of government were responsible for various aspects of the project resulted in a lack of coordination in dealing with many issues that, inevitably, arose. A coordinating committee would have been valuable, but there wasn't one. As the project moved toward its end, in 1979, the government did launch a coordinated study on fisheries development; the study report showed concern about the slow progress of this project.

The government determination to develop fisheries, combined with the presence of an able private sector, produced an improvement in the quality of government staff and services. Government has continued to support training and fishing harbour development, although the main national investment agency has become sufficiently disillusioned with fisheries development that it is not at present supporting it.

The international funding agency had intended to support construction of an industrial fishing port with full facilities after the harbour study was completed. However, after the study dragged on for, altogether, 14 years (and at one point recommended the site of an existing port that was then evidently developing by itself) the agency decided that rising costs had made the development uneconomic.

At appraisal an economic rate of return for the project vessels had been estimated. However, conditions changed so much during the project, and the parts of the program that were completed encountered such different operational conditions to what had been envisaged that no comparative economic evaluation was practical.

The present economic situation and the lack of a fully coordinated fisheries administration continue to impede the full development of fisheries in this country. None the less, progress is being made in private fishing operations.

The school will increase its impact with a residential dormitory for students from all parts of the country, greater cooperation with the private sector, and field services and extension courses using the greater supplies of instructional materials recently provided.

Meanwhile there have been increases both in the number of industrial fishing boats based in the country and in the number of foreign vessels fishing the country's territorial waters. Some 60 processing, canning, or fishmeal plants have been set up, and there has been an increase in production from cultivated shrimp ponds. These improvements have stemmed from private investment as well as bilateral aid.

Selected references

Anon. 1976a. Estado actual de la educación y capacitación en ciencias del mar en el Ecuador par Instituto Nacional de Pesca, Guayaquil, Ecuador. Seminario Regional Valparaiso (Chile). Rev. Com. Perm. Pac. Sur., 6, 49–62.

1976b. Política nacional sobre educacion pesquera en Ecuador – Dir. Gen. de Pesca. Guayaquil. Ecuador. Rev. Comm. Perm. Pac. Sur., 6, 43–47.

1978. Bank backs a giant project. Fishing News International, 17(12), 35.

Artunduaga, P. E. 1978. Consideraciones sobre el nucleo de pescadores de Malaga en el Pacifico Colombiano. Bogota, Colombia, Divulgación Pesquería, Instituto Desarrollo Recursos Naturales Renovables, 13(2). 14 p.

Brownell, W. 1978. Extension training of artisanal fishermen and other fisheries personnel in the WECAF region. Rome, Italy, FAO, WECAF Report 19.

Bullis, H. R., Klima, F. 1972. The marine fisheries of Panama. Bulletin of the Biological Society of Washington, 167, 78.

Cole, R. C. 1976. Fishery training needs in WECAF countries. Rome, Italy, FAO, WECAF Report 1.

Day, C. 1979. Most of the fishing will stay small in WECAF. Fishing News International, 18(2), 48–49.

FAO (Food and Agriculture Organization). 1977–79. Fishery country profiles for Belize, Colombia, Costa Rica, Ecuador, El Salvador, Guatemala, Honduras, Mexico, Nicaragua, Panama, Peru, Venezuela. Rome, Italy, FAO, compiled for WECAF.

1980. Interregional fisheries development and management programme (WECAF component), Interim Report. Rome, Italy, FAO, WECAF Report 3. 53 p.

Guidicelli, M. 1979a. Aspectos Técnicos de la Pesca Artesanal en la República Dominicana y recomendaciones para su mejoramiento y desarrollo. Rome, Italy, FAO. WECAF Report 5.

1979b. La Pesca Artesanal Marítima en la Costa Caribeña de Colombia – situación, posibilidades y necesidades para el desarrollo. Rome, Italy, FAO, WECAF Report 8. 50 p.

1979c. Programa de desarrollo de la pesca artesanal en la region de San Andrés y Providencia, Colombia. Rome, Italy, FAO, WECAF Report 25. 23 p.

Guidicelli, M., Wirth, A. J. 1979. Informe de la Mision a Hondruas con relación a diversas posibilidades de desarrollo pesquero en el pais. Rome, Italy, FAO, WECAF Report 26.

Klima, E. F. 1976. A review of the fishery resources in the Western Central Atlantic. Rome, Italy, FAO, WECAF Study 3/E. 77 p.

Pollnac, R. B. 1977. Panamanian small-scale fishermen: society culture and change. Kingston, R.I., USA, University of Rhode Island, ICMRD, Maritime Technical Report 44. 91 p.

Roedel, P. M., Saila, S., *eds*. 1979. Stock assessment for tropical small-scale fisheries. Proceedings of an international workshop. Kingston, R. I., USA, University of Rhode Island, ICMRD. 237 p.

Sutinen, J. G., Pollnac, R. B. 1979. Small-scale fisheries in Central America – acquiring information for decision making. Kingston, R.I., USA, University of Rhode Island, ICMRD. 612 p.

6
Fisheries credit for marine fishing and aquaculture in Southeast Asia

Background

This chapter examines a large-scale project in Asia where an international lending agency financed a considerable development involving fishing vessels, fish carriers, shore facilities, technical assistance, as well as fishpond construction and rehabilitation.

The Asian country had a large population and extensive territorial fishing grounds. At the time the project was proposed, total fish production was somewhat more than one million tonnes, giving a per-capita consumption estimated at 28 kg annually. Thirty-seven percent of fish landings came from substantial industrial operations by more than 2 000 licensed vessels, and the main production (63%) was contributed by a widespread artisanal sector that included inshore fishing and extensive areas of coastal and inland aquaculture.

Fish accounted for half the animal protein eaten in the country. Fisheries activities occupied 6% of the total labour force – about 750 000 workers – and contributed 4% of the country's gross domestic product. The country annually exported nearly $6 million worth of shrimp and tuna but imported more than $20 million worth of fish, mostly canned mackerel from Japan and fishmeal from Peru.

The industrial fleet was restricted to waters deeper than 12 metres. These vessels used a wide variety of fishing methods, though vessels larger than 70 tonnes were mainly trawlers and purse seiners. By 1970 trawlers were showing better financial returns and thus were on the increase. However, this changed after the project was in effect owing to fuel price increases.

There were no canneries. Attempts to establish them had failed, because production costs of the canned products were not competitive with international prices, and it was cheaper to import Japanese canned mackerel as there was no tariff protection. Accordingly, other than the exports of frozen tuna and shrimp mentioned above, the fish catch was consumed domestically. Most fish was sold fresh, generally without ice, and consumed daily. Anything left over would be sold at a discount for drying and smoking and subsequent sale inland, where there were no cold-storage marketing facilities.

Shore services generally were inadequate. There was no established, organized fishing port, although studies to that end were under way. The geography of the country did not lend itself to the establishment of one large, central fishing port, but there was a clear need for a network of wholesale storage, distribution, and retail facilities. Ice-making, cold storage, and refrigerated transport were inadequate. Accordingly, any temporary increase in supplies strained the existing storage and depressed market prices.

Vessels could not readily be serviced. The country had more than 30 shipyards, but few facilities for slipping and repair of boats or maintenance of deck machinery and electronic equipment.

Available data were inadequate for planning the long-term development of the fisheries. However, it was estimated that the potential catch

Fig 15 Malaysian fish trawlers at anchor in harbour

from the territorial open seas was about 1.6 million tonnes, which would allow for a 150% expansion of current industrial type operations. Artisanal operations in coastal and inland waters were apparently exploiting some 80% of the resources available to them, but doing so without any backup of services, credit, or marketing arrangements. The 170 000 hectares of ponds were producing more than 100 000 tonnes of fish, a substantial contribution. The yield per hectare was, none the less, well below that achieved in other nearby countries. No scientific determination had been made as to the best size of pond, whether in family-type operations or in the increasingly popular, larger commercial operations that used hired labour. Only about 6 000 hectares of freshwater ponds were in operation, inasmuch as fish-farmers preferred to produce shrimp and other high-value species from brackish-water ponds.

Objectives

In 1973 an external lending agency signed an agreement with the national government to assist fisheries development. Funds would be advanced, through the national development bank and with the assistance of the department of fisheries, to individuals and companies to finance the acquisition of vessels, fish carriers, and equipment and the installation of slipways and icemaking plants. There would be technical advice and studies for marine fishing systems and for aquaculture, rehabilitation of ponds damaged by recent typhoons, and construction of new ponds. Over four years, the project would provide for:

- Construction of 60 new vessels. Of these, 15 were to be steel-hulled, 130 tonne boats, 28 metres long with diesel engines of 750 horsepower, costing $180 000 each. They would carry crews of 17. The remaining 45 vessels would be wooden, 24 metres long, 70 tonnes, with 336 horsepower engines, crews of 15, costing $59 000 each. Both types of vessel were intended for trawling and would be fully equipped with radio, fishfinders, and mechanical winches. The boats would be built locally after expatriate technical staff had set specifications for improved design of vessels. The equipment was to be largely imported free of import taxes. The project included provision for replacing equipment on 100 older vessels. These would get winches, power blocks, generators, transceivers, echo-sounders, and fishfinders at an average cost per vessel of about $7 500.

- Purchase of fish transporter vessels. These would be boats of less than 75 tonnes, bought secondhand abroad, to carry the catch from distant fishing grounds to market areas. Each carrier would be able to service two to three fishing boats, according to season. Expatriate staff would draw up tender documents and specifications for the vessels and equipment.
- Three icemaking plants. Each plant would cost about $200 000 and produce 30 tonnes daily, providing sufficient crushed ice for 20 vessels, some part of the existing fleet and some new ones.
- Two slipways. These would be located in rural communities for the repair of fishing boats. With marine railways they would be able to service vessels up to 150 tonnes.
- Rehabilitation of flood-damaged brackish-water fishponds. This was to cover some 4 500 hectares of fishponds at a cost of about $300 per hectare.
- Reconstruction of fishponds. This would involve improvement of the layouts and fish-rearing practices in existing brackish-water ponds, at a cost per hectare of about $740.
- Construction of new, freshwater fishponds. One hundred hectares were scheduled, at more than $1 000 per hectare.

Fig 16 Fishing with bright lights used less fuel in Philippines

Technical assistance included training for the appraisal staff of the national development bank and for 100 key extension personnel by experts in aquaculture with support from the department of fisheries. About six senior personnel were to receive training abroad. A consultant firm would study the feasibility of small fishponds and survey fish marketing.

Local costs of the project were projected at $10.15 million and foreign-exchange costs would be $8.35 million, including a provision for inflation. The host government asked the lending agency for $11.6 million, the full amount of the foreign-exchange costs and half the inland fisheries component costs. The agency agreed and lent the money on a 17-year repayment schedule with a four-year grace period. The marine part of the project was expected to realize a return of 22 to 34%, and the inland fisheries were expected to return 23 to 27% on the investment during the five-year period.

The national development bank would administer the loans to sub-borrowers, with help from the department of fisheries. This bank had already had considerable experience in fisheries loans, but for this project it decided to reorganize its fisheries lending arrangements. The lending appraisal and loan supervision operations were separated. Another change was in the requirement for collateral; previously the bank had required security such as real estate, and this had precluded many possible borrowers. Now the loans could be largely secured against the boats, which had to be fully insured. The credit program and the changes were widely publicized. The bank set up a section to evaluate periodically the impact of the program. It would also assess previous loan policies and review procedures for future operations.

The project was expected to create 1 000 new, permanent jobs in the extra fishing vessels and icemaking plants and another 1 000 for family members or hired labour in the fishponds. In addition, there would be an improvement in the diet mostly of low-income and rural people. The responsibilities to be undertaken within the national bank and the department of fisheries would impart greater expertise and have an institution-building effect.

Inputs

Work began on the inland part of the project more or less on signature in 1973. The marine part became effective about six months later, but the lack of sufficient trained staff at the national development bank also

Fig 17 Fishcarrier vessels bring the catch to wholesale centres

caused a slight delay. The result was that the procurement of the vessels and equipment coincided with the rapid inflation consequent on the 1973–74 world oil crisis. The funds allocated thus were inadequate to buy all the vessels and equipment planned. For instance the 70–tonne wooden vessels originally estimated at $60 000 would cost, by 1976, $220 000. The cost of diesel fuel quadrupled and similar increases occurred in the price of fishing nets made of petroleum-based materials.

Accordingly the project funds permitted only 35 vessels, rather than the 60 first planned. The 130–tonne steel design was reassessed and seen to be no longer economic; thus all 35 vessels constructed were of wood, which itself increased in price and became scarcer between 1974 and 1976. There were some variations in design, and those built were of 72 to 100 tonnes. The design changes resulted in further delays and higher costs.

The sharply higher fuel costs affected the economics of fishing. Trawling is fuel-intensive, and by 1979 fuel accounted for 55% of the cost of operating a trawler. Accordingly the demand for purse seiners increased, and in fact 12 of the 35 boats constructed were purse seiners. To save fuel, vessels operated closer to base in more heavily exploited areas. Trawlers operated fewer days and the fleet failed to increase its catch from the more distant, less-exploited areas, as had been planned.

The plan to acquire 10 fish carriers was abandoned for three reasons: Few available boats met the criteria; the procurement procedure was too slow to acquire boats overseas when they became available; and the high cost of steel meant that there was a greater international demand for used boats, which resulted in something of a sellers' market.

Only one 10-tonne-capacity ice-making plant was constructed, against three 30-tonne plants proposed. The one plant built proceeded slowly because of construction delays and difficult procurement of equipment. However, private entrepreneurs were also at that time building ice plants near project locations.

Outputs

Only one of the two proposed slipways was built, and that one functioned at only 50% capacity. Demand for slipway services was lower than expected because of their high cost, and vessel owners curtailed expenditures due to higher operating costs consequent upon the fuel crisis.

The allocation for replacing equipment on older vessels was not used. Many of the older boats examined were not worth the expense of upgrading their equipment for a limited operational period.

The bank provided nearly 700 loans to sub-borrowers for rehabilitation of fishponds, covering about 1 000 hectares more than had been planned. One reason for this extra interest in fishponds was the great increase in the seagoing costs of fish production. The fishpond developments cost 24% more than had been envisaged, but the increase in production was 58% higher than expected. All the development was in brackishwater ponds; none of the freshwater pond construction was carried out.

Of the land developed as fishponds, 78% was in enterprises of more than 20 hectares. Annual production in the new ponds averaged 1.9 tonnes per hectare, against 0.6 t/ha throughout the country. Fishponds between 20 and 50 hectares showed the highest yields; rehabilitated ponds showed higher yields than new ones; owner-operated ponds showed higher yields than those under hired management. Better production resulted where there was formal training of pond operators, who applied improved techniques using fertilizers and supplementary feeds.

The national bank broadened its expertise in loan appraisal and supervision and in financing vessels.

The advisory services originally proposed were replaced by bilateral technical aid. This had a favourable effect on the country's extension services. The government fisheries service, in the inland part of the program, conducted several training courses using five main extension units. The service developed an excellent training program that produced a number of highly qualified extension personnel and teaching aids.

Effects

The need to conserve fuel led to shifts of emphasis in the program – the abandonment of steel hulls, use of purse seiners rather than trawlers, and operations closer to shore, thus eliminating the need for fish carriers. Another effect was the greater interest in artisanal fishing using manual systems. Similarly, fishponds became more important as a less fuel-intensive means of production.

The country's boat-building capabilities improved through the guidelines introduced by consultants for use by the bank. The 12 purse seiners that were built instead of trawlers employed crews of 30 to 35 each, against the 15 normally aboard trawlers. Fuel accounted for 45% of their operating costs, against 55% for trawlers, and the species they caught were of higher value. Thus purse seining was more profitable, especially with the modification and adoption locally of new techniques such as the use of floating, attracting devices.

During the four years the project was operating, there was a 150% increase in the wholesale price of fish; the rise in production costs, however, was ever greater.

The improved expertise within the national development bank occurred both at headquarters and at its many branches. The bank trained its field staff and prepared a manual to guide them in fishery loan procedures. A useful innovation, which has been continued, was to organize courses for the borrowers. Regional and international agencies, including FAO/UNDP, also helped in national and international training activities for the borrowers.

The fishermen and fishpond operators who borrowed money found the terms onerous; in particular the penalty clauses on late repayment were considered burdensome. The government bank was somewhat more rigid than the private banks, although its interest rate was lower; it mostly provided loans for long-term investment. By contrast, the private banks aimed mainly at the short-term loan market; borrowers found their loan procedures quicker, and such banks usually asked for less information before granting the loan. The latter circumstances were preferred by private entrepreneurs who wanted quick loans to capitalize on good seasons and preferred not to disclose too many business details to government.

The role of credit in fishing and fishpond operation was clearly becoming more important, and bank investment in fisheries was on the increase, although it remained less than that in agriculture. By the end of the project about 17% of private bank loans was in fisheries, compared to 35% in agriculture. This growth of credit appears not to have spread to the poorer artisanal fishermen, but small groups of them are gradually being helped by the national bank. Faster loan processing and a more enterprising approach to risk by the bank would be helpful in future.

Despite the more painstaking loan procedures by the national bank, the quantity of the loans it made was lower than those of the private sector, and only 30% of inland loans and 25% of marine loans were repaid within the assigned period.

Difficulties

The yearly typhoon disasters and the fluctuations in market prices posed hazards for fishpond operators, and they felt that a system of crop insurance and a distress fund would be valuable. Borrowers felt that

Fig 18 Coastal aquaculture is widespread in S E Asia

interest rates and penalty clauses should be reviewed and that the national bank should increase its extension efforts to enhance fishpond management efficiency and new systems of fishing, perhaps in regular collaboration with other technical organizations. As a means to ensure profitable investment in current operations and as a management measure, the limitation of licences for fishing vessels in certain localities should also be considered.

Artisanal fishermen complained that they had to pay taxes on gear and fuel because they bought these at local retail stores, while larger operations, which get import permits, did not. The extension to them of these tax-free concessions, possibly through cooperatives, would lower the costs of producing fish from coastal waters and fishponds.

Another difficulty was the high cost of vessels and gear. This forced the fishermen to think of using second-hand equipment. But imports of second-hand vessels and gear would likely run into problems of servicing, although the general ability of the country to maintain mechanical equipment was good.

Many entrepreneur producers pointed out the need to stabilize the market price of fish. Associated with this was the need for adequate storage and education of consumers to accept frozen fish and a wider variety of fish products. Fish could be used in a variety of attractive products convenient for consumer use such as fish cakes, fish balls, and fish sausages, which are made from various types of fish. Currently the packaging and use of deboned fish were still in the initial stages in this country. However, stable consumer products that would be suitable for rural households without domestic refrigeration were sorely needed. Fishermen believed that development of such products would help stabilize their markets, and they favoured support by the national bank for investment in facilities to manufacture them.

Benefits

This project was sufficiently successful that it led to another one budgeted at $12 million, then to a third and fourth. Investment in fisheries increased. In 1976 the government bank supplied $24.5 million and private banks supplied $102 million. In 1981 the government bank was supplying $15 million and the private banks were supplying $249 million.

The collaboration between the national bank and the department of fisheries which started with this project continued. Eventually fisheries departmental staff were seconded to the bank for permanent liaison, thus contributing to improved services.

Impact

Although the specific impact that had been planned did not come about, the increased area of fishponds, their greater yields, and the 35 new vessels contributed to higher production. In addition to the one slipway and one ice-making plant that the project financed, the private sector funded the other facilities that had been envisaged. No purchase of fish carriers or improvements to older vessels were undertaken. The increase in fishpond production of about 19 000 tonnes attributed to the project was about twice what had been planned. Conversely, the 8 900–tonne increase in marine production was only 59% of that expected. In both sectors the project introduced and helped establish new fishing and fishpond production systems.

The additional employment sought was almost entirely achieved in the inland part of the project. In the marine sector, the purse seiners built had crews of 30 to 35 each, compared to the expected crews of 15 for wooden trawlers and 17 for steel trawlers. Though this factor to some extent offset the decrease in numbers of vessels put into service, only about half the anticipated employment in marine fisheries was actually generated.

Local builders assimilated the improvements in vessel construction practices that the project introduced. Owners and operators of the new vessels appreciated the improvements in design, despite the indifferent financial performance that resulted from external circumstances beyond their control.

The research aspects of the project contributed a better knowledge of the fishing grounds, enabling new and continuing measures to be taken to prevent over-exploitation. The marketing study, though not resulting in new projects, provided data that the country has already been able to use. The training aspects had a considerable beneficial impact.

Evaluation and assessment

On the whole, the international lending agency was satisfied with the performance of the national development bank and felt it had made the best choice of banks to administer the program. The conditions laid down for this project emphasized that the national bank would improve its service to fisheries, but there might also have been more effective conditions calling for continued collaboration with the department of fisheries. Some parts of the loan agreement, such as acquiring the fish carriers, ice-making plants, and slipways, could not be carried out, wholly or in part, while consultancy services for vessel design were deemed to be satisfactory. The international agency was able to modify the agreement by raising the loan limits to accommodate local inflation. It developed an effective, continuing relationship with the national bank, which enhanced the bank's machinery for administering subsequent loans.

The national bank would like to have seen more of the reports on which the lending agency made its decisions. National staff felt that had they seen the detailed appraisal report, rather than just the final loan document, there would have been a basis for reviewing the rationale of local policy decisions. Likewise, they felt that had they been shown the critical supervisory reports to the international agency they might better have been able to appreciate the operating policies and procedures of the agency and thus improve the performance of their own systems.

One change in policy that would have been appreciated by the government fishery administration was to have assistance for training programs as grants rather than loans that required repayment. These could perhaps have come through FAO or UNDP. Certainly, for efficient performance, loan recipients should have received more training in operational techniques.

National bank staff also believed that grants, rather than loans, would be more appropriate for studies on fish stock assessment and economic utilization systems for available fish resources. Much of this basic information pertains to the national and shared resources of the region; that is, the common property fish stocks. They felt that grant-aided research here should be available to all countries in the region that share the stock for mutual benefit in its effective management. Mission-oriented research of this type needed a higher priority and timeliness than it receives under existing arrangements, such as through the UNDP.

Overall results

The national bank prepared a number of detailed analyses of the operations of various types and sizes of vessels and compared the costs and benefits of various methods of operation. Such studies are exceedingly valuable in investment planning, but to be comprehensive they need more details of the vessel operations of the private sector. Operators have traditionally been secretive about this type of information, but if it can be obtained – say on the understanding that individual operations are not identified publicly – it would be an essential tool in fisheries development and investment.

Loans secured by real estate collateral showed a better recovery rate than those not so secured and based largely on vessels and equipment covered by insurance. However, the program was aimed at medium and small enterprises that often cannot produce adequate immovable property collateral. Lower interest rates are needed to produce better loan performance in these circumstances. In general, risks and collateral in fishing loans are a world-wide problem in which the difficulties of lending operations have to be balanced against the need for the development. Financing for fisheries development cannot be regarded in the same light as ordinary commercial lending – a field in which existing private banks already have elaborate and widespread services. The role of government development banks should be to help establish activities that cannot as yet be supported by commercial banks according to commercial criteria.

Fishermen thus look to development banks for long-term capital on favourable terms. More intensified training for the borrowers should improve their performance.

In this project, some emphasis shifted to fishpond culture from marine fishing. This was partly because the fuel crisis imposed different methods of working on the fleet, as previously described, and reduced the economic rate of return on marine fishing. Also, fishpond systems are more like agriculture; the real-estate collateral is clearly available and assessment and supervision are easier. The operatives felt crop insurance to be more necessary than with rice or other agricultural products; typhoons or heavy rains can result in a total loss of the current crop through disappearance of the fish into natural bodies of water, whereas with planted crops there remains the possibility of recovery, and, if not, there is the tangible evidence of crop failure.

The experience gained in this project was applied to subsequent projects, which were more effectively coordinated and supervised thereby. The national bank prepared more complete final reports on subsequent projects, objectively analysing them. In retrospect, therefore, much has been learnt in this country that can be applied to development projects elsewhere.

Fig 19 Fish pens and enclosures are well developed in the Philippines

Selected references

ADB (Asian Development Bank). 1981. Fisheries Subsector. Manila, Philippines, ADB Agriculture and Rural Development Department. 66 p.

Anon. 1981. Integrated fisheries development plan for the 1980s. Manila, Philippines, Ministry of Natural Resources Fishery Industry Development Council.

Campleman, G. 1973. The transition from small-scale to large-scale fisheries industry. Journal of the Fisheries Research Board of Canada, 30(12,2), 2159–2165.

Charusombat, V. 1979. Development of marine resources in Thailand. Thai Fisheries Gazette, 32(1), 5–12.

Cole, R. C. 1973. Fisheries development and requirements of fishery education and training in Malaysia, Thailand, Fiji and the Philippines with particular reference to the artisanal sector. Rome, Italy, FAO, FAO Fisheries Report R143/E, 58 p.

Day, C. 1977. A report on conditions and needs of marine small scale fisheries in South Asia. Colombo (Sri Lanka). FAO/UNDP project for development of small-scale fisheries. Journal of the Inland Fisheries Society of India, 10, 9–18.

Emmerson, D. K. 1980. Rethinking artisanal fisheries development: Western concepts, Asian experiences. Washington, D.C., USA, World Bank, Staff working paper 423. 97 p.

FAO (Food and Agriculture Organization). 1976. Report of the BFAR/SCSP workshop on the fishery resources of the Visayan and Sibuyan area, Tigbauan, Iloilo,

Philippines, 18–22 October, 1976. Manila, Philippines, SCSP, SCS/GEN/76/7. 26 p.

1977. Report on the BFAR/SCSP workshop on fishery resources of the Sulu Sea and Moro Gulf areas, Cagayan de Oro, Philippines, 25–29 April, 1977. Manila, Philippines, SCSP, SCS/GEN/77/11. 58 p.

1978. Report of the BFAR/SCSP workshop on the fisheries resources of the Pacific coast of the Philippines, Naga City, Philippines, 18–22 September, 1978. Manila, Philippines, SCSP, SCS/GEN/78/19. 48 p.

1979a. Report of the workshop on the tuna resources of Indonesian and Philippine waters, Jakarta, Indonesia, 20–23 March, 1979. Manila, Philippines, SCSP, SCS/GEN/79/21. 35 p.

1979b. Report of the BFAR/SCSP workshop on fisheries resources of the north Luzon and western coasts of Luzon, 18–20 April, 1979, Manila, Philippines. Manila, Philippines, SCSP SCS/GEN/79/22. 57 p.

1980. Symposium on the development and management of small-scale fisheries, in Proceedings, 19th session, Indo-Pacific Fisheries Commission, Kyoto, Japan, 1979. Bangkok, Thailand, FAO.

Firth, R. W. 1966. Malay fishermen, their peasant economy. Hamden, Connecticut, USA, Archon Books. 398 p.

Hongskul, V., Chullasorn, S. 1979. Marine fisheries resources and management problems in the ASEAN region. Thai Fisheries Gazette, 32(3), 285–290.

Joseph, K. M. 1977. Economics of 38' GRP fishing vessels issued by the ADB fisheries project. A case study. Symposium on development of offshore and deepsea fishing, Bulletin Fisheries Estuarine Station, Colombo, 28, 100–108.

Kwale, W. L. 1980. Long-line catches of tuna within the 200-mile economic zones of Indian and Western Pacific oceans. Rome, Italy, FAO, IOFC/DEV/80/48.

Lawson, R. M. 1972. Report on credit for artisanal fishermen in Southeast Asia. Rome, Italy, FAO, FAO Fisheries Report R122/E. 63 p.

Long, S. W. 1973. The South China Sea Fisheries Aquaculture Development: Status, potential and development of coastal aquaculture in the countries bordering the South China Sea. Manila, Philippines, SCSP, SCS/DEV/73/5. 51 p.

Lisac, H. 1979. Some technical aspects of small-scale fish landing facilities. Manila, Philippines, SCSP, SCS/79/WP/81. 32 p.

Liv, H. C., Kao, C. L. 1979. General review of demersal fish resources around Taiwan. Acta Oceanographica, 9, 77–96.

Lockwood, B., Ruddle, K., *eds*. 1977. Small-scale fisheries development. Social science contribution. Honolulu, USA, East–West Centre. 215 p.

Marr, J. C. 1976. Fishery and resource management in Southeast Asia. Washington, D.C., USA, Resources for the Future, Study 7.

Mendis, A. S. 1978. Demersal fishery resources of Sri Lanka and its present level of exploitation in off-shore and deep-sea waters. Bulletin Fisheries Resources Station Colombo, 28, 47–51.

Munro, G. R., Chee Kim Loy. 1978. The economics of fishing and the developing world. A Malaysian case study. Penang, Malaysia, University Sains. 139 p. plus supplements.

Pannayotov, T. 1980. Economic conditions and prospects of small-scale fishermen in Thailand. Marine Policy, April 1980, 142–146.

Payne, R. L. 1972. Planning criteria for large-scale fisheries development with special reference to the Indian Ocean in FAO technical conference on fishing development and management, Vancouver, 1979. Journal of the Fisheries Research Board of Canada, 30, 2321–2327.
Phasuk, B. 1979. Status of the pelagic fisheries in the Gulf of Thailand. Thai Fisheries Gazette, 32(1), 99–106.
Sakiyama, T. 1981. Long-term trends and problems of fisheries development in the South China Sea – a case of fish catch economy in the countries of humid tropics. Presented to 12th Pacific Trade Conference, Institute of Development Economics, Tokyo, Japan, 1981.
Sardjono, I. 1980. Problems and strategies in the development of fisheries in Indonesia. Proceedings of Oceanexpo, Oceantropique, Bordeaux, France. Paris, France, Technoexpo.
Senanayake, N. 1978. Infrastructure facilities for deep-sea and off-shore fishing. Bulletin Fisheries Resources Station, Colombo, 28, 92–99.
Sharp, G. D. 1979. Areas of potentially successful exploitation of tunas in the Indian Ocean with emphasis on surface methods. Rome, Italy, FAO, IOP/DEV/79/47. 55 p.
USAID (US Agency for International Development). 1977. Fisheries sector study of the Philippines – Report of USAID/Philippines fisheries mission. Manila, Philippines, USAID. 79 p.
Vijayan, U. K. 1978. Comparative efficiencies of mechanized fishing crafts introduced in India. Symposium Report. Bulletin Fisheries Resources Station, Colombo, 28, 71–76.
Wickramasinghe, U. K. 1978. Role of state and financial institutions in financing the fishing industry. Symposium report. Bulletin Fisheries Resources Station, Colombo, 28, 110–111.

7
Institutional organization for inland fisheries research and development

Background

Our sixth case history, the last that concerns one specific country, is a project funded by a bilateral foreign aid agency to upgrade research and development capability in an Asian country with a population of more than 30 million.

This country, somewhat more than 20 years ago, had taken the decision to improve its fishing capabilities. Coastal waters had abundant resources, and consequently the first development plan had concentrated on marine fishing and had established an increasingly important industry. The marine fishery produced substantial amounts of fish for consumption and export. None the less, inland fisheries were in some ways more important for the interior population and provided the main component of a balanced diet for inland residents.

In 1979 total fish production was 1.7 million tonnes, of which inland fisheries contributed only 200 000. Much of the latter was subsistence fishing; fishing for food was the constant practice in rivers, canals, and all available bodies of water in the country. The catch of fish was consumed fresh and processed in many forms. The impact of the increasing production of marine fisheries was less pronounced inland, because the cost of transporting the fish put its price beyond the reach of inland consumers, while the quality of marine fish also deteriorates the further it is transported. The main source of inland fish consumed by poor rural communities was that caught directly by families, as this involved no cash expenditure. Thus the swamps, lakes, rivers, and reservoir impoundments were important sources of inland fish supplies.

Official estimates of per capita consumption ranged from 9 to 27 kg yearly in different parts of the country. Overall consumption was about 20 kg – equal to 47% of the total animal protein consumed. Some 43% of the 20 kg came from freshwater fisheries. Agricultural development adversely affected inland fish production when large flood-plain areas that previously had supplied more than half the inland fish were drained for crops.

In response to a request for assistance for inland fisheries development in 1972, a bilateral aid agency sent an identification mission, which recognized a need for comprehensive staff training to improve biological investigations that addressed national priorities. There were many small fish-culture stations that had little impact on inland fisheries management and could not conduct applied research. There were several areas of overlapping interest, and a clear need existed to organize the most economic use of available facilities, equipment, and services and to provide for those essential elements that were inadequate for fishery development activities.

Accordingly the mission proposed the establishment of a centre for inland fisheries research and development, to be closely associated with the main national university. This institute would involve present staff, students, and facilities, using the existing network of field stations and extension service to disseminate research results. This would ensure the long-term upgrading of formal instruction and applied research capability of university students. The institute would also do field research, during which students would learn while interacting with field workers.

The project was planned to begin in 1972 and continue for five years. Staff from the donor country's institutions provided short-term consultation and long-term expertise, and there was a twinning arrangement between the new centre and a major applied research institution of the donor country. Because of the considerable environmental and climatic difference between recipient and donor countries, senior personnel from the developed country provided general guidance and laid down research plans that the new centre might follow. These were carried out jointly by local and expatriate staff rather than directed by external experts, as had previously been the operational procedure.

Objectives

Objectives of the project were to conduct applied research on inland fisheries management, to promote practical measures to restore and protect inland fish habitats, to set up a research and study centre to guide the host government, and to provide coordinated training programs for staff of the centre in association with the country's agricultural university.

The program first aimed to establish the laboratory buildings and installations for field experiments, while training nationals at post-graduate levels for biological investigations. Library facilities were to be improved; donor and recipient personnel were to select appropriate scientific equipment for required facilities and laboratories.

The national government provided land, buildings, furniture, and some equipment, built experimental ponds, and paid most local operating costs. The donor government provided equipment and paid the travel and accommodation costs of expatriate personnel and some operating costs.

The donor budgeted $400 000 for staff services, $130 000 for 26 man-years of training of country nationals (two doctorates and six masters' degrees), $100 000 for library services, including publications, $250 000 for scientific equipment, and $120 000 for operating and contingency funds for a total of $1 000 000. The host government budgeted $350 000 in buildings and furniture, $100 000 in land, pond construction, and other preparation, $180 000 for operational expenses, $100 000 for miscellaneous and support services, and $20 000 for contingencies. This amounted to $750 000, which was increased on review to $1 000 000.

The first year of the program allowed for planning of the facilities and preparation of the host country's budget. Expatriate and local staff defined targets and stipulated the equipment needed. The detailed formulation of the project had to be completed within six months, so the host government could arrange its contribution.

In the second year construction of buildings and field installations began and students to be sent abroad were identified. Expatriate personnel also undertook local on-the-job training. During this time there began the planning and operation of research projects; studies in taxonomy, parasitology, and environmental factors were to begin when trained national personnel became available. Library facilities were improved, and donor and recipient together selected appropriate equipment.

Several research projects investigated the rehabilitation, management, and protection of available natural waters, as well as fish habitat improvement in altered natural waters such as river systems, irrigation reservoirs, and hydroelectric water bodies. The work was also to include studies for optimizing the fish-culture systems widespread in the country.

Inputs

At the end of the implementation period, staff comprised 46 biologists, of whom 18 were PhDs, and a support staff of about 205. The donor agency provided six senior short-term consultants and five longer-term consultants. Twenty-one people were sent on staff training programs; of these seven were PhDs and 13 of them were sent on specific observational missions. The institute has a current strength of 225 staff members, of whom 50 are biologists.

An impressive building to house laboratories and administrative services was erected on the university campus. There were laboratories for chemistry, biology, pollution, pathology, and nutrition investigations plus aquarium, library, lecture room, and photographic and drafting facilities. The building cost more than $2 million in 1976 and would likely have cost nearer $6 million in 1983. The building included a model system of electrical, air, and water supply, which was new for the country and permitted easy control and servicing. The chemical and pollution laboratories had modern equipment of a standard previously

Fig 20 Aquaculture production increase contributes significantly to national fish requirements

unknown in the region. Working conditions were excellent and national staff who had received training abroad found the technical facilities very satisfactory, a circumstance that is said to have enhanced their performance and their professional standing in the region.

The support services given by both governments involved seem to have followed the agreement, although local circumstances caused some delays. Construction and equipping of the buildings presented the biggest difficulty. The library became well-established, with 4 000 listings and 2 000 texts. There are 80 journal subscriptions and an inter-library loan service. The library has photocopying service and planned to have an on-line computer data link with FAO information systems.

The institute set up liaison with local fishery centres within the country and with inland fisheries centres in other countries within the region, including the 20 field stations. The institute has provided previously unavailable services to universities, government, and international fishery agencies; researchers have used its accumulation of reference literature. The library still needed a microfiche reader and printer, a valuable tool in the humid tropics for document storage and one that would facilitate the exchange of literature with other libraries.

Fig 21 Researchers supervise university student training

Outputs
During the project implementation period, 23 training programs and 14 conferences were organized. The feed mill produced 57 tonnes of fish feed. There were 16 training workshops provided by fishery technicians of the institute, and 372 fishery projects were completed. Production of juvenile fish from the hatchery facilities totalled 15.1 million of 10 different species; 11.3 million were distributed to aquaculture stations, the balance being placed in public waters. Six new special projects were started in rural communities. The institute prepared 18 new instructional papers for farmers, of which 29 000 copies were distributed. There were also three exhibitions, two demonstrations, and nine filmed TV programs on fisheries, which were carried on several local stations, and there were many articles in the local and national press.

Some of the services rendered are shown in this table:

	1975	*1977–78*	*1981*
Fishermen's visits to institute	739	2 392	3 271
Technical services rendered	551	691	2 065
Extension pamphlets distributed	5 000	17 000	12 700
Library acquisitions		2 792	1 732
Annual operational budget ($)	90 000		497 000

In 1981 the program provided 11 training programs for national staff, produced 9.2 million seed fish, of which 1.5 million went to aquaculture and 7.7 million to public waters, produced 32 tonnes of fish feeds, undertook 27 research projects, and completed 36 field studies.

Effects
The institute has achieved a high profile and the head of state of the country has given unprecedented recognition to its contribution to improving fishing conditions in rural communities. It also enjoys regional and international prestige. It has attracted aid from donors and international investment agencies that have not previously supported fisheries projects in that country. The institute has provided the basis for national policy decisions on inland fisheries and their relationship with national fisheries. The trainees, fish, feed-formulae, and services of the institution are used throughout the country, and its work is no longer performed wholly at the institute but has spread through many stations

in the country. Senior biologists of the fishery stations spent time working with institution staff, producing a more effective interaction between the remote field stations and central administration and providing feedback, which affected future policy decisions.

During project implementation there was an increase in staff numbers, but later the general economic crisis caused the government to impose a freeze on staff recruitment. None the less, the increasing expertise of staff has produced a growth in capability and confidence, and staff members are publishing more technical papers in local and regional journals. The progress in developing staff seemed slow at first, but the pace quickened cumulatively.

The host country considered the training part of the project to be particularly useful, although it has not been fully adequate to meet continuing needs. None the less, this part of the program has considerably strengthened staff capabilities, with doctoral researchers especially providing useful service. Short-term training abroad was considered the weakest part of the program as all trainees needed to stay longer to adjust to new conditions, and the training was not as well focused as it could have been. The observation visits for senior personnel were valuable, as they helped broaden their understanding of how to manage such an institution. These visits were both technically educational and stimulating for their administrative decisions and policy perspectives.

However not everything went perfectly for the laboratory and central buildings of the institute. There were some technical problems; a spare generator arrived too late to be hooked into the system during construction, and although the air conditioning is appreciated, the cost of fuel has made it a high-expense item owing to recent inflationary trends. The cost of necessary electric services absorbs much of the maintenance and operational budget. Some of the imported equipment is difficult to service, parts being hard to get, and not all of it is ideally suited to local conditions.

The reputation of the institute is now established nationally and regionally. Its national activities are the basis for technical development, planning, and operational activities. It is used by multilateral and bilateral agencies as the logical regional centre for inland fisheries research training, embracing not only aquaculture, but also other inland fisheries, notably reservoir and river fisheries. There are plans for a regional training centre, a fish genetic bank, and a breeding hormone bank. Disciplinary units have been established for investigating fisheries management, population studies, ecology, biology, pollution, aquaculture, fish nutrition and health, taxonomy, and extension services. There have been four technical studies on aquaculture systems, 11 technical surveys of reservoirs and river basins, plus 39 other ongoing studies, and institute staff by the end of the project had published 72 papers. The government has recognized the value of the institute and has accordingly increased its operational budget to $0.5 million in 1982, exclusive of staff salaries. The annual departmental budget was $6.9 million.

Banks and other investors have used institute staff for appraisals and advice on development projects.

While visitors to the institute were increasing from 300 a year to more than 3 000, outreach activities and relationships with university students were improving. Students receive instruction within the institute, but also as part of their studies, do field work in various field stations that are managed by personnel previously at the institute.

The institute has promoted those species of fish that are the most profitable for farmers. Some of the institute's studies have been aimed at maximizing fishpond production or controlling disease or other problems that the farmers have encountered with particular fish species. When changing environmental or market conditions raised problems, the institute formulated research programs to resolve them.

Impact

The institute has regular dealings with staff of universities and industrial services in the immediate vicinity of the capital city as well as with banks, regional river commissions, development programs, and the personnel of UNDP, FAO, and other international agencies.

The 20 inland fisheries stations form a network throughout the country in which trainees and ex-staff members of the institute conduct training programs for farmers. These rural extension services provide the university students who work in them with a better understanding of, and orientation in their professional activities. At three field stations staff urged the need for greater support to enable more effective transfer of newly developed technology. More public waters – reservoirs, canals, and dams – are now being stocked with fish to provide food for rural communities. These new resources are in part replacement of the food fish formerly available from wild stocks in the flood plains. There is a greater need for management of the new stocks.

When the institute was in its early stages, the central administration was perhaps reluctant to allow it fully to carry out its original mandate. However after the institute had established a 'track record of achievements,' it evolved as a main factor influencing development of fisheries. It now addresses both research and teaching as well as planning and development of fisheries. The country's marine fisheries have limited scope for further development and accordingly the inland fisheries are becoming more important to satisfy food requirements in the economic and social development of inland communities. Indeed, given that more than half the marine catch has been used for livestock feeds (pigs, poultry, fish, and prawns) more fish as food for inland rural people is a clear national priority.

Evaluation and assessment

It took longer to establish the institution than had originally been envisaged. Possibly the timetables for construction, equipment, and staffing were not realistic. However, the project is now established and has settled into routine activities, as the performance data show. It is important to recognize that the original idea, of setting up a nucleus institution, has been broadened. Today the institution and its outlying components serve the needs of the country's fisheries and contribute, through academic activity, to fishery policy and planning as well.

Whereas donor staff became available quickly, the process of assembling and training national staff took somewhat longer than anticipated. Some of the installations and equipment are not yet functioning as envisaged. Nevertheless, the major installations are completely functional; with what is available a viable institution has been functioning and effectively serving centrally and throughout the interior areas of the country.

The donor agency was somewhat less than flexible when changes in timing and financing became advisable. Looking back, however, it can be seen that it put only $1 million into the project during five years, whereas the host country in the end spent about $6 million (which would have been $10 million in 1983 dollars). The donor agency pressed the host country to live up to its commitments, even at increasing cost, but itself seemed at first reluctant to accede to requests for more consultant staff time, supervision, or training. When the donor had met its obligations the staff of the institute endured a period of uncertainty, wondering whether any further support would be forthcoming. At least this made

Fig 22 Experiments show significant results for increased fish yields in ponds

them more reliant on their own government. In the event, other donor agencies, recognizing that a worthwhile institution had been created, came forward with further support.

The equipment, although chosen jointly by national and expatriate staff, was not always best suited to local conditions. Differences in electrical cycles were initially a problem. The cost of operating electrical equipment had not been fully allowed for. There were instances when the equipment might better have been constructed locally or within the region.

The institution clearly needed more reference literature to make it adequately self-sufficient. More visits from senior consultants may have been valuable and appropriate during later phases of implementation.

There remained a need for extension training, including visits to other countries with similar tropical environments. Perhaps some type of special fund for this purpose needs to be established, as the training funds allocated to the institution by its government tend to be for other specific local purposes. The donor agency agreed to allow doctoral students to do their course work in the donor country but their research in the host country. This was a useful precedent, thoroughly endorsed by those students who followed this procedure. They have thus been able to continue their contributions to solving local problems and provided experience study material for formal instruction. Further, the standards and level of their research have been validated by the more prestigious overseas university.

During the project the host country provided two project managers and the donor country three project leaders. Although the host country was slow in carrying out its obligations, it was not appreciably slower than would be normal there. The staff allocated by the host government to the institution represented an increase in the manpower of the fisheries department, and this increase continued for a while after the end of the project until the government froze all new recruitment as a matter of national policy. The long-term trainees found their most valuable instruction was received overseas. For greater effectiveness, short-term training needed to be better focused; however, local training continues and the training that has already been carried out can be shown to be valuable in the field. The planning of research projects, formerly done on an *ad hoc* basis, is now more coordinated. Thus, problems are listed as requiring a solution in either three months, six months, or a year or more. Improved planning of research projects, including an assessment of economic benefits, has also resulted through a jointly coordinated system of the institute.

The institute has had an important national influence, besides attracting attention and support regionally and internationally. Some of the published studies have resulted in new policies – for instance, one study of a reservoir in the northeast showed that fish generated more revenue than the electricity produced by the hydroelectric plant for which the dam had been particularly constructed. The government thereupon decided to build four similar dams with a total impoundment of nearly 4 000 km^2 and more than 100 smaller ponds, each about 1 km^2, that would not only conserve water for irrigation but, more important, provide food on site for the surrounding communities.

The training programs of the institute are being extended to coordinate with local efforts throughout the country. Neighbouring countries also are using these programs as a pattern, often with the help of former institute staff.

The economic benefits from the institution's activities are not easily estimated. There are those who add up the value of fish from new reservoirs or new fish farms, while others contend this new production would have largely come about anyway; certainly there seems to have been a very real improvement on upland aquaculture.

Follow-up projects are being supported by ADB, FAO, IDRC, JICA, UNDP, and the World Bank. These projects, amounting to more than $100 million, cover fish culture, hydroelectric dams, fisheries in swamplands, lakes, and rivers, units for fish health, pollution control, and extension services. It is contended that this would never have happened in the absence of such a prestigious institute.

The institute's research staff believed that without foreign influence in this project, the government would not have set up such an institution and services, even though they were clearly needed. However, having committed itself to the institution, the government was obviously impressed with its work and successive administrations have continued to support the development of inland fisheries. The institute's effectiveness has attracted local and foreign aid; it now serves as a vehicle for development and the focus of new programs.

The country now is able to cope with changing circumstances, notably those brought about by the new EEZ. Previously this country had a marine fishing fleet of 20 000 vessels, which fished not only their

Fig 23 Trapping fish on their spawning migration in fast flowing rivers requires effective controls

own waters but the areas off the coasts of neighbouring countries. Promulgation of the EEZ had the effect of restricting these operations largely to the country's own marine fisheries, which resulted in overfishing and diminishing returns. This in turn created greater pressure for fish from the inland fisheries. Fortunately, this research institute was being brought into being just as the EEZ was introduced, and it gave significant help in planning for inland development. Now there are some 625 inland fisheries personnel, of which 325 are in field stations. Most of them have received instruction and guidance through the institution.

The government has given greater priority to the local publication of papers to satisfy the pragmatic extension needs of the country and less priority to those that make contributions to international journals of scientific knowledge. Thus the number of publications may not seem impressive by international standards but may adequately satisfy local needs. But there is a clear need to upgrade the local instructional manuals and to stimulate international scientific exchange through the journals. It would be valuable if present plans for regional journals and manuals could be coordinated through international agencies. Contributions to present extension manuals by experienced regional experts would enhance their value to farmers and improve techniques in the field, for this country and its neighbours.

More than 100 pesticides have been introduced in recent years into southeast Asia. These all have different effects, sometimes a debilitating effect, on fish that live in natural waters and are exposed to the chemicals through natural drainage. Some of this water is also pumped into fishponds. It is of continual importance to study the varied toxicology of these pesticides. Further staff training in this new and difficult subject is needed. The debilitation of fish through pesticide pollution results in a lower survival rate for young fish. Any adverse climatic or ecologic change may induce a greater susceptibility to parasites or diseases, and there have been significant financial losses, which have been directly attributable to bacterial diseases. Problems have been experienced with catfish culture and even wild stocks. Personnel of the institute have studied these conditions.

The impact of the institute on fisheries development has clearly been considerable. It has been an investment in far-sighted thinking that is now beginning to bear fruit beyond the expectations of those who planned it.

Selected references

Bati, A. 1977. Aquaculture development in Bangladesh, in Proceedings, 17th session, Indo-Pacific Fisheries Council, 1977, Bangkok, Thailand, IPFC 17, Part 3, 189–194.

Bhukaswan, T. 1980. Management of Asian reservoir fisheries. Rome, Italy, FAO, FAO Fisheries Technical Paper T207/E. 69 p.

CIFRI (Central Inland Fisheries Research Institute). 1976. First 138 case studies of composite fish culture in India. Barrackpore, West Bengal, India, CIFRI, Bulletin 23. 138 p.

1979. Final report of CIFRI/IDRC rural aquaculture project. Barrackpore, West Bengal, India, CIFRI. 1250 p.

Dussart, R. H. 1974. Biology of inland waters in tropical Asia, in Natural resources of humid tropical Asia. Paris, France, UNESCO. 331–353.

Eusebio, J. A. 1977. SEAFDEC fisheries research management and programs, in Proceedings, agricultural research workshop/seminar, Kuala Lumpur, Malaysia, 1977. Las Banos, Philippines, SEARCA. 11 p.

FAO (Food and Agriculture Organization). 1976. Aquaculture planning in Asia. Report of the Regional Workshop on Aquaculture Planning, Bangkok, Thailand, 1–17 October, 1975. Rome, Italy, FAO, ADCP/REP/76/2/E. 154 p.

1980. Comparative studies on freshwater fisheries: Pallanza workshop report, 1979. Rome, Italy, FAO, FAO Fisheries Technical Paper 181. 42 p.

George, P. C., Sinha, V. R. P. 1975. A ten-year aquaculture development plan for India, 1975–1984. Rome, Italy, FAO, FAO/UNDP ADCP.

Guerrero, R. D. 1976. Aquaculture research commodity in the Philippines. College Laguna. PCARR Fisheries Resources Commodity Workshop, 1976. Las Banos, Philippines. PCARR. 50 p.

1977. The national aquaculture research programme of the Philippines. Resources System Research Congress, Mindanao, Philippines, 5 p.

Jhingran, V. G., Tripathi, S. D. 1977. National perspective of inland fisheries of India, in Proceedings, 17th session, Indo-Pacific Fisheries Council, 1977. Bangkok, Thailand, IFPC 17, Part 3, 41–58.

Karim, M. 1979. Suggestions for inland fisheries development of Bangladesh. (Planning Commission, Dacca, Bangladesh.) Farm Economy, 1, 323.

Kloke, C. W., Potaros, M. 1975. Aquaculture as an integral part of the agricultural farming system. Case study in Northeast Thailand. Bangkok, Thailand, IPFC. IPFC occasional paper 1975/4, 10 p.

Kunkle, S. H., James, J. L., *eds*. 1977. Guidelines for watershed management. Rome, Italy, FAO, FAO Forestry Department Conservation Guide 1. 293 p.

Librero, A. R. *et al.* 1976. Survey of the aquaculture industry in the Philippines. Las Banos, Philippines, SEAFDEC/PCARR research paper series 1. 57 p.

Long, S. W. 1973. Review of the status and problems of coastal aquaculture in the Indo-Pacific region, in Coastal aquaculture, IPFC Symposium, 1973. Bangkok, Thailand, IPFC/FAO.

Lowe-McConnel, R. H. 1975. Fish communities in tropical freshwater. New York, USA, Longman. 337 p.

Mendis, A. S. 1977. The role of man-made lakes in the development of freshwater fisheries of Sri Lanka, in Proceedings, 17th session, Indo-Pacific Fisheries Council, 1977. Bangkok, Thailand, IPFC 17, Part 3, 247–257.

Mori, S., Ikusima, I., *eds.* 1980. Proceedings, workshop on promoting limnology in the developing countries, 1979. Tokyo, Japan, JICA. 172 p.

Nasaruddin, A. 1979. Aquaculture as a supplement to land development schemes in Malaysia, in Proceedings, Conference on natural resource management in developing countries, 1979. MARDI, Serdang, Malaysia.

Pantulu, V. R. 1976. Role of aquaculture in water resource development – a case study of the Lower Mekong Basin Project. Rome, Italy, FAO. FAO Kyoto Conference Paper FIR AQ. Con/76/E20.

Pillay, T. V. R., *ed.* 1972. Coastal aquaculture in the Indo-Pacific region, London, England, Fishing News Books. 497 p.

Rabanal, H. R. 1974. The potentials of aquaculture development in the Indo-Pacific region. IPFC Working Party. Bangkok, Thailand. IPFC. 34 p.

Ricker, W. E., *ed.* 1971. Methods for assessment of fish production in fresh waters. IBP Handbook no. 3. Oxford, England, Blackwell.

Saraya, A. 1978. Aquaculture is compatible with coastal mangrove management in Thailand, in Department of Fisheries annual technical report 1978. Bangkok, Thailand, Department of Fisheries aquaculture survey section. 11–23.

Silas, E. G., *ed.* 1981. Proceedings of seminar on the role of small-scale fisheries and coastal aquaculture in integrated rural development. Cochin, India, Central Marine Fisheries Research Institute Bulletin 30-A. 203 p.

Sumawidjaja, K., Muluk, C. *et al.* 1977. Fishery ecological survey in Indonesian watersheds; aspects of fish protection in public waters. Bogor, Indonesia, IPB, Inland Fisheries Department. 82 p.

Suwingo, P. 1976. Reservoir fisheries and the analysis of their production increase. Bogor, Indonesia, SEAMEO Regional Centre for Tropical Biology BIOTRIOP. 9 p.

Tapiador, D. P. *et al.* 1977. Freshwater fisheries and aquaculture in China – Report of the FAO aquaculture mission to China. Rome, Italy, FAO, FAO Fisheries Technical Paper 168. 84 p.

8
Bilateral fishery inputs into a multilateral program of UNDP

Background

The next project for discussion is a multilateral program that had inputs from bilateral aid agencies. The South China Sea Fisheries Development and Coordination Program (SCSP) was originally advocated by the Indo-Pacific Fisheries Council (IPFC). Its activities were supported by UNDP and executed by FAO. It involved countries bordering the South China Sea – Hong Kong, Indonesia, Kampuchea, Malaysia, the Philippines, Thailand, and Vietnam. After considerable discussion, the program decided to concentrate development efforts on four aspects of fisheries: pelagic, demersal, crustacean and molluscan resources, and aquaculture. There were also studies and workshops on stock assessment, statistics, exploratory fishing, economic analysis, legal and institutional requirements, service facilities, marketing and processing, boat operations, and small-scale fisheries. The Canadian International Development Agency (CIDA) was asked to support the pelagic resources survey and agreed to do so. This chapter reviews this component of SCSP and related activities.

The SCSP program began in 1972 as a well designed series of projects. In the pelagic fisheries part of the program, for instance, there was semi-commercial test fishing for the most important species, plus feasibility studies for new and ongoing investment in vessels, and analyses of the profitability of exploiting available resources. It should be appreciated that at that time developing countries were anxious for quick results from fisheries; thus specific data that would make investment attractive to venture capital were in demand. This information included, for pelagic fisheries development, data on seasonal distribution and abundance, size of schools, fish behaviour, catch rates, and the most efficient design for boats for the existing operations.

It seemed that the pelagic fish resource surveys that the UNDP/FAO had done elsewhere were less than acceptable to governments. Even the recent approach of using acoustic surveys for species determination as a first step to fishery development, was not acceptable to governments. They needed results with quick commercial application, to satisfy the food demands of their increasing populations, by extending the operations of their existing national fleets to harvest underutilized or unexploited resources.

Accordingly, in this program the decision was made to use chartered commercial fishing boats and crews with the best available equipment for exploratory and test fishing. A supporting acoustic search program would determine the catch rates of unexploited sources of marketable species. CIDA's funding of this practical program was to cost $2.8 million, starting in 1972 and continuing for 3½ years. The survey was to be conducted with two commercial purse seiners and a spotter aircraft and include feasibility studies and crew training programs.

Inputs

The costs were: purse seiners, $1 million each; aircraft charter, $200 000; additional equipment and fishing gear, $50 000; personnel services, $280 000; travel, $75 000; and project management fees, $170 000, with $35 000 for contingencies. Participating countries supplied local operating costs, including personnel services, port facilities, and some equipment.

As can be readily appreciated, it was not possible to separate the pelagic part of the program wholly from the other aspects, and by integration with other aspects of the whole SCSP, it had a ripple effect. The CIDA activity conducted 30 surveys, published 25 cruise reports, held several training sessions, and harvested pelagic fish that sold for $1.593 million, which went back into the SCS program. The target area was, however, too vast to be covered in the allotted time, although the surveys did increase the available knowledge of pelagic fish in various parts of the area.

Outputs

The surveys produced evidence of commercially fishable resources in Philippine waters; vessels carried out surveys throughout the year. These showed where tuna was available year-round, indicated some seasonal fluctuations, and established catch rates for various seasons. The surveys helped to improve the ability of countries to assess fish stocks and to train nationals for implementing regional resource management.

The assessments were used by the Asian Development Bank to evaluate an investment for pelagic fisheries development in Thailand. Small and medium-sized pelagic stocks were identified and catch rates established in continental shelf and coastal waters in the Andaman Sea off Malaysia. The CIDA project demonstrated test-fishing with slightly different gear and light attraction on dark nights, and this system was perfected by Philippine entrepreneurs. The system is now in widespread commercial use by the Philippine fishing fleet. The tuna located and thus harvested were previously unexploited, so this was a notable gain.

The CIDA project was not able to cover comprehensively all the areas off Thailand, nor did time permit full coverage off Hong Kong and Indonesia, but for all the operations that were undertaken it published survey reports. In general, the surveys and test fishing were as comprehensive as time and other constraints permitted. The aerial survey was not fully implemented, so within the time available it was not possible to ascertain seasonal patterns of migration of pelagic fish throughout the program's operating area. The test-fishing vessels did not have access to all territorial waters, especially those of Indonesia.

The SCSP sponsored workshops on methods of planning, coordinating, and evaluating surveys of pelagic resources. In addition to furnishing this methodology to fishery personnel of participating countries, the program provided all the equipment needed for the detection and assessment of fish stocks. Initially there was a problem with the availability of personnel trained to use and maintain this equipment, but eventually it was possible to complete surveys and apply the results promptly.

The proceeds from the catches of the test-fishing vessels amounted to a sixth of the cost of the whole SCSP. Significantly, this helped to support the continued activities of SCSP when additional funds were needed. The catches were also used for test marketing and experimental canning studies for domestic and export markets. The test marketing showed patterns of consumer acceptance and resistance for various fish products and preservation methods, and, as a result, some products test-marketed are now being sold by private industry.

The project largely achieved its aim in pelagic fishing and achieved some of its aims in the feasibility studies and training programs. These various activities continued and were refined after the end of the project inputs. Bearing in mind that the CIDA support project was a component of the whole project, it is important to indicate the other aspects.

The SCSP sponsored 16 training courses, 38 study tours, and 38 workshops or seminars on a wide variety of subjects for fishing personnel within the region. The program issued 115 working papers, of which 23 were on pelagic resources, held 46 workshops or seminars, of which 13 concerned pelagic fisheries, published five instructional manuals, and issued 16 periodic reports and 30 technical papers.

The workshops, training courses, and study tours involved 852 participants from 18 countries from 1974 to 1982. Most of these were from the Philippines, where the program is based, but Indonesia, Thailand, and Malaysia were also well represented.

Effects

The demonstration by the test-fishing vessels in Philippine waters had an evident effect. Local fishermen adapted the methods to their vessels, improved on the system of floating attracting devices, and eventually established fleets to harvest the tuna. By 1978 eight fishing companies operated 50 purse seiners, landing 10 000 tonnes a year, and by 1982 these operations were yielding $25 million annually. Subsequently the use of floating attracting devices spread throughout the western Pacific. Other promising stocks of pelagic fish located elsewhere were not, however, exploited.

However, artisanal fishermen's associations voiced resentment over these two foreign purse seiners, notably at regional meetings of FAO-IPFC. Their opposition also resulted in adverse press comment in donor countries over the effect of the pelagic survey on inshore fishermen.

The root of the objection contained in their leaflets was that large-scale fishing investments led to "overexploitation of the resource, depressed markets and caused the collapse of precariously impoverished fishing communities, unable to protect themselves and only capable of earning a living from the sea." They contended that traditional fisheries are more labour-intensive but are "in harmony with the sea's productivity, thus ecologically safe." Sophisticated, mechanized, purse seiners or trawlers, they argued, "deplete the resource and deprive the small fishermen of their fishing grounds, and thus threaten the survival of coastal villages, benefiting only a few nationals with greater returns to foreign industry, and being contrary to the best interests of the local economy and social stability." This caused a reassessment of the project and a reviews of policies for the whole SCSP. This incident had very

Fig 24 Investment in fishing fleets results in competitive fishing for resources shared by many countries

significant ramifications for small-scale fisheries development policy, worldwide.

Thus greater emphasis was subsequently laid on coastal fisheries, aquaculture, and small-scale fishing. The test marketing of landed catches from the pelagic survey was discreetly directed to channels that were not competitive with small-scale landings, and there was less publicity for the Philippines development of tuna fishing. Some participating countries found it inexpedient to permit access of the vessels and spotter plane into their waters. The survey was not extended. CIDA subsequently supported fish processing and small-scale fisheries development in Thailand, the Philippines, Malaysia, and Indonesia.

In general, it was evident that activities in regional pelagic and demersal fisheries increased considerably, as did organizational, management, training, and aquaculture activities. The SCSP encouraged subsequent projects in each of the participating countries, as well as further bilateral support, all of which seems to have been effectively coordinated. Thus the program had an impact on government policy and administration, industrial fishing, and artisanal fishermen. With regard to its general activities, no previous regional program had covered so broad a range of activities or had stimulated or was involved in so much preparation and development. The program was seen as catalyst, organizer, and coordinator, although other agencies were responsible for follow-up. Program staff found themselves assisting other missions in fisheries development in the region.

The SCSP undertook a number of research and promotional activities, some of which were financed by the $1.6 million generated by the sale of fish caught by the survey vessels.

During the pelagic surveys, more than 12 technical feasibility studies and other economic assessments were conducted. In Malaysia and the Philippines socioeconomic studies were also undertaken, one being a model demonstration project, and four being studies of fishing communities. The related activities of the SCSP used these other studies to plan the integrated development of artisanal fishing. Studies of this type provided the basic information for the Asian Development Bank loan project of $27 million to Malaysia to upgrade the shore facilities that service small fishing communities.

In the Philippines, studies helped establish the priority programs to deal (with local fishing problems with specific suggestions for improving current technology and upgrading shore facilities) to assist small-scale fisheries and fishermen's organizations. Supporting studies and workshops on statistics, legal, and institutional aspects, marketing, and other topics have produced further useful information. Reviews were undertaken on the collection, compilation, and dissemination of data in member countries. Review teams provided advice and support to the fisheries departments of various member countries, including the planning and development of their statistical services.

Legal and institutional studies took place initially in Malaysia and the Philippines, and later in other countries. They seemed useful in identifying needs and constraints in present legislation. The recommendations of many of these studies have been accepted and put into law by the countries concerned.

The Asian Development Bank financed market studies under the program. A number of national fisheries profiles were published. There were studies on resource potential and exploitation levels, and prospects for development. The ADB used such studies as the basis for its fisheries loan programs. Participating countries also have followed the studies with their own programs.

The SCSP was flexible and clearly met the needs and wishes of member countries. Some of the studies were in response to their requests; others were under the auspices of the ADB or were part of the appraisal process. Yet more studies were part of the UNDP small-scale fisheries development activities in individual countries. Many *ad hoc* studies responded to the pressing needs of national governments to find ways of reducing costs, under the pressure of inflation and the fuel crisis.

The major impact of SCSP studies has been to provide background and intelligence that the countries in that region could use for investment and development. They have also led to further studies by regional and national banks. The program thus has been a significant catalyst. Economic and marketing studies have been the start of continued gathering of information within the region to improve fisheries. Published studies have helped member countries improve their organizations and regulatory structures.

Although no formal analysis of training needs was published during the period reviewed here, there were institutional and on-the-job training workshops, plus some on-the-job training on the fishing vessels. It is hard to measure the value of these exercises, but certainly this type of technical information exchange and exposure to management strategies had not previously been so intensively presented on so many different aspects of fisheries problems. The bringing together of regional personnel was valuable in clarifying problems that are universal throughout the region and in enabling all these personnel to understand each other's viewpoints. Crucial issues could therefore be mutually addressed, and neighbouring countries were enabled to work together to share their resources.

The project was evaluated through annual tripartite reviews, a midterm review, and a final assessment. Numerous progress reports were compiled every six months, as were reports of the project coordinating committee. Thus all parts of the program were regularly documented. Continued impact of the program may depend on the effective use of these documents.

Impact

It was evident in the workshops and other activities that the SCSP management was effective and that national, regional, and international fisheries agencies collaborated well. Government fisheries departments also worked well with other agencies. Personnel of the investment banks involved had close liaison with other organizations and were able to use the data they collected in making their investment decisions. The UNDP-FAO management clearly did a praiseworthy job in channelling information to where it was needed.

The pelagic fisheries survey was both a practical demonstration with concrete results and an encouragement to member countries to follow it up. Being part of the overall program, it was integrated with other activities, notably training, and the revenue brought in by the catch funded further activities. With this pattern established, six donor countries provided similar bilateral aid within the program for eight other sub-projects.

Even after the CIDA support came to an end, the project's effects remained. The results and published reports guided national fisheries departments in investing in catching, landing, processing, storage, and marketing of fish; international banks used the project's experience in making their investment decisions, and the project provided some trained people for new fishing techniques. Another after-effect was that the number of fishing boats and the amount of fishing gear patterned after the CIDA exploratory vessels' specifications continued to increase. Indeed the self-priming activities of entrepreneurs was one of the most important results, particularly such activities as the adaptation of purse seining techniques with floating attracting devices in the Philippines.

Governments were particularly interested in the practical exploitation of their pelagic fisheries, and government follow-up depended on the financial and administrative abilities of their fisheries departments. Studies on joint ventures legal and administrative structures proved useful to them.

The improvement in fisheries knowledge and techniques was hastened by changes in circumstances during the program. There was a stronger emphasis on small-scale fisheries and on trained manpower; the fuel crisis made purse seining more economic than trawling. Accordingly, the program served to stimulate these developments. International investment funds (of a project originally intended for trawling) were changed to buy purse seiners to harvest the newly discovered pelagic fish.

Much remains to be done. Not all countries bordering the South China Sea have taken up the lessons learnt, and the work that has been started could continue at a higher technical level of skill. Effective statistical monitoring should conserve the newly developed stocks of pelagic fish, as well as those in the inshore areas traditionally fished.

The UNDP ended funding for the program as such in 1983, after 11 years, but a permanent committee to manage the South China Sea fisheries succeeded it. The IPFC in 1982 reported that besides the $15.5 million actually spent by the program and its bilateral inputs, further foreign-assisted fisheries projects to the tune of $35 million had been implemented, and $200 million was invested by the World Bank and the Asian Development Bank. All this activity may not have directly arisen from the SCSP, but all the projects used the information and expertise generated during the carrying out of the program.

Overall regional impact

The preceding paragraphs show the extent of activities of this 'umbrella-type' program. Member countries appreciated particularly that many of the activities could be put in hand quickly in response to need, whereas the usual aid program does not get under way until after lengthy procedures. One highly placed government official observed that it is now up to governments and the private sector to maintain the momentum of development from aggregated experience in the region, while avoiding over-investment or other counterproductive practices.

The most impressive practical development occurred in the Philippine purse seining for medium pelagic fish, mainly frigate and bullet tuna. The success of a few entrepreneurs in modifying their methods triggered an upsurge in production from 26 200 tonnes in 1973 to 96 874 tonnes in 1980.

Perhaps the main lesson learnt from the South China Seas Program has been that, although several projects within the program have succeeded in their aim, continuity is not thereby assured. The national fishery personnel have acquired the competence and confidence to continue their work; now that the program is formally ended they must look elsewhere for the money to do so. Hopefully, their own governments, having recognized the benefits of such inter-country consultation, will find money to support continued management programs for their valuable common-property resource.

It is important that regional projects of this nature be planned so that the impact may continue even after the project itself ends. Provisions to ensure this continuity may include:

- National lead centres specializing in particular disciplines should be selected according to carefully established criteria, and they should be developed as centres for the analysis of technical data in collaboration with other such centres within the region.

- Research, management, and training institutions should be so strengthened that they can collaborate with others in the region on the most urgent fisheries needs.
- Technical working parties on fishing operations, resource management, regulations, training, processing, and marketing should continue to get operational funds.
- There should be regular meetings within the region for the exchange of technical information and the harmonizing of policies and legislation on shared fishing resources.

These provisions would lead to continued collaboration in fisheries development such as is now provided by fisheries commissions in the North Pacific. Spin-offs might include jointly funded manuals and specialized services. As personnel in these centres acquire expertise, their stature and that of their institutions will increase. Provisions of this nature should be pursued during the course of projects such as the SCSP, because obviously, even where the project is successful, continued action will eventually be in the hands of national institutions.

The benefits of an SCSP-type program will not, one hopes, disappear

Fig 25 The wide variety of fish products marketed

with the ending of the program, but there needs to be a constituted, international body to manage and coordinate shared fisheries. The form it takes and the way it is sponsored is less important than having an authoritative technical staff, free of political interference, to carry out rational management and ensure the survival of an intensive fishery.

Lessons learnt

A number of lessons may be taken from the experience of the South China Seas Program:

- Surveys should not be expected to cover an unrealistically large area and there should be time to accomplish the tasks effectively. The most effective promotions are those within a small area, where the population is receptive and a demonstration project can be shown to signal a profitable commercial venture. For instance, the program might support receptive entrepreneurs and publicize their success. This is better than gathering a collection of data that may not have immediate commercial application, while ecological and economic circumstances are constantly changing.
- The program dramatized the social and political importance of small-scale fisheries.
- Policy planners need to be assured that regional management of the fisheries is well established and that the program has produced mechanisms for continued operations, for authoritative analysis of resource problems, and for regular technical consultations between the experts of participating countries. Ways of doing this permanently, but more quickly, should be explored.
- There is an urgent need during operations to strengthen national institutions so they can become the leading regional centres and provide personnel for critical tasks.
- The program management must relate closely to those in the fisheries industry so as to get feedback.
- National and regional banks must be involved in training programs. The program management must provide data on which the banks can make good investment appraisals.
- The program showed the benefits of early involvement of policymakers, investment institutions, and fishermen rather than relying for advice only on technical fisheries staff to promote development subsequently.

- Governments seemed to believe that international or regional organizations should be able to respond promptly to problems that arise in fisheries management. This ability has become of greater significance with the promulgation of the EEZ.
- During this program regional bodies such as the Southeast Asian Fisheries Development Centre and the Association of South East Asian Nations have evolved. They may be the logical vehicles for future UNDP and bilateral inputs for managing the development of shared resources. In other regions there may be parallels.
- In the SCSP, participating countries were well informed of what was being done by numerous reports. It would be useful to know which of these reports were most urgently needed for development (and what publications were considered most useful by their target audiences). An assessment of the value of such reports to users in each country might be enlightening. The evaluation could consider the audience, type of report or manual, its detail and content, and what should be revised and adapted or what new texts or manuals might be needed in similar programs.
- The practical commercial fishery for pelagic fish that evolved in this program may be a model for other areas with shared resources and limited staff and funds.
- The program's management of tropical, multispecies fish stocks and its coordination of legal and institutional measures may also be a model.

Selected references

Periodic progress reports: Sixteen were issued by the SCSP from 1974 to 1981 under document numbers SCS/PR/74/1 to SCS/PR/82/16, author Woodland, A. G. The full reference for the first of these is as follows:

Woodland, A. G. 1974. Project progress reports of the South China Sea Fisheries Development and Coordinating Programme. Manila, Philippines, SCSP, SCS/PR/74/1. 19 p.

Coordinating committee reports: Reports of the *ad hoc* coordinating committee meetings of the South China Sea Fisheries Development and Coordinating Programme were issued from 1974 to 1979 under document numbers SCSP: 74/1 REP to SCSP: 79/8 REP.

Survey vessel cruise reports of 25 exploratory test fishing cruises by two chartered vessels, *Southward Ho* and *Royal Venture*, by 10 authors in 15 reports were issued from 1975 to 1978, under document numbers SCS/75/WP/8 to SCS/78/WP/78. These reports were also summarized in:

Chikumi, S., Simpson, A. C., Murdoch, W. R. 1978. Test fishing for tuna and small pelagic species by FAO chartered purse seiners in Philippine and South China Sea waters 1974–1977. Manila, Philippines, SCSP, SCS/DEV/78/18. 87 p.

See also:

Boucher, J. M., Hinds, L., Lang, E., Wong, E. 1976. Canadian contribution to the SCSP evaluation report. Ottawa, Canada, CIDA. 133 p.

Fisheries technical papers

Doucet, F. J. *et al.* 1973. Institutional Legal Aspects: Institutional legal aspects affecting fishery development in selected countries bordering the South China Sea. Manila, Philippines, SCSP, SCS/DEV/73/9. 32 p.

FAO (Food and Agriculture Organization). 1974. Species identification for fishery purposes. East Indian Ocean (fishing area 57) and West Central Pacific (area 71). 4 Vols. Rome, Italy, FAO.

Kume, S. 1973. Tuna resources in the South China Sea. Manila, Philippines, SCSP, SCS/DEV/73/4. 18 p.

Menasveta, D. *et al.* 1973. The SCSF Pelagic Resources: Pelagic fisheries resources of South China Sea and prospects for their development. Manila, Philippines, SCSP, SCS/DEV/73/6. 63 p.

Mistakadis, M. N. 1973. The SCSF Crustacean Resources: The crustacean resources and related fisheries in the countries bordering the South China Sea. Manila, Philippines, SCSP, SCS/DEV/73/7. 39 p.

Pope, J. 1979. Stock assessment in multispecies fisheries with reference to Gulf of Thailand trawl fishery. Manila, Philippines, SCSP, SCS/DEV/79/19. 106 p.

Ruckes, E. 1973. SCS Fisheries Marketing and Trade: Fish utilization, marketing and trade in countries bordering the South China Sea. Manila, Philippines, SCSP, SCS/DEV/73/8. 8 p.

SCSP (South China Seas Fisheries Development and Coordinating Programme). 1976a. Development potential of selected fishery products in the regional member countries of Asian Development Bank. Manila, Philippines, SCSP, SCS/DEV/76/11. 107 p.

1976b. Fishery country profiles. Manila, Philippines, SCSP, SCS/DEV/76/11 appendix 1. 173 p.

Thompson, D. B. 1079a. Marine fisheries extension. Presented to ASEAN seminar/workshop on fisheries extension, Manila, Philippines, 18–25 February. 41 p.

1979b. Training requirements for the fisheries of Southeast Asia. Presented to SEAFDEC consultative meeting on fisheries training, Bangkok, Thailand, 14–18 May. 11 p.

Workshop reports

SCSP (South China Seas Fisheries Development and Coordinating Programme). 1974. Report of the workshop on planning and coordination of resources survey and evaluation in the South China Sea, 28 August – 4 September, Manila, Philippines. Manila, Philippines, SCSP, SCS/GEN/74/1. 197 p.

1976. Report of workshop on legal and institutional aspects of fishery resources, management and development, 5–8 April, Manila, Philippines. Manila, Philippines, SCSP, SCS/GEN/76/3. 95 p.

1977. Report on the training workshop on fisheries statistics, Malaysia, 12–21 October, 1976. Manila, Philippines, SCSP, SCS/GEN/77/10. 27 p.

1979a. Report of the workshop on the tuna resources of Indonesia and Philippine waters, 20–23 March, Jakarta, Indonesia. Manila, Philippines, SCSP, SCS/GEN/79/21. 35 p.

1979b. Report of the consultation meeting on management of tuna resources of the Indian and Pacific oceans, 26–29 June, Manila, Philippines. Manila, Phillipines, SCSP, SCS/GEN/79/24. 155 p.

1980. Report of the workshop on application and results of acoustic methods for resource appraisal surveys in the South China Sea, Manila, Philippines. Manila, Philippines. SCSP, SCS/GEN/80/25. 19 p. Annexes 1 to 4.

1981. Report of the regional seminar on monitoring, control and surveillance of fisheries in exclusive economic zones, Jakarta, Indonesia, 30 November – 4 December. Manila, Philippines, SCSP, SCS/GEN/81/31.

Working papers

Cleaver, W. D. 1975. A preliminary design and general arrangements for an offshore purse seine vessel for the East Coast of West Malaysia. Manila, Philippines, SCSP, SCS/75/WP/18. 35 p.

Moore, G. 1978. Legal and institutional aspects of fisheries management and development – a second interim report. Manila, Philippines, SCSP, SCS/78/WP/78. 23 p.

Wheeland, H. A. 1976. Statistics for fisheries development. Manila, Philippines, SCSP, SCS/76/WP/52. 11 p.

Other references

Aprieto, V. L. 1980. Philippine tuna fisheries resource and industry. Fish. Res. J. Philippines, 5(1), 53–66. Reprinted Honolulu, USA, East–West Environment and Policy Institute 19. 14 p.

Ouchi, K. Y. M. 1980. Japanese Fisheries in Southeast Asian Seas under the New Law of the Sea Regime. Reprinted Honolulu, USA, East–West Environment and Policy Institute 17. 16 p.

Valencia, M. J. 1978. S. E. Asia: National marine interests and marine regionalism from ocean development and international law, Journal of Marine Affairs, 5(4), 421–476.

9
CIDA bilateral inputs into CECAF–UNDP inter-regional project

Background

The area of the Eastern Central Atlantic from Gibraltar to the mouth of the Congo River contains the most important fishery resources off the west coast of Africa. Before the late 1950s, fishing along the coasts of the region consisted mainly of inshore operations by non-powered canoes. The subsequent entry of foreign fleets into West African waters, plus increasing mechanization of fishing vessels by coastal countries, have led to heavy exploitation of some of the stocks of the area.

Fish is an important economic resource to many West African countries. It is a source of employment, income, foreign exchange, and food. Per-capita consumption of fish in a number of West African countries exceeds by several times the world average of 12 kg/yr, and even in countries of low consumption it is a food appreciated by a wide cross-section of the population. The total population of the region is 180 million, and the annual fish production is 3.5 million tonnes, which includes the production of coastal and non-African states. The northern countries, with about 25% of the population, face 75% of the fish stocks, which are, in general, heavily exploited with large catches being taken by non-coastal states. In the southern part of the area, say south of latitude 10°N, there is 75% of the population facing 25% of the regional fish resources. These facts pose multi-faceted social economic and political issues.

Despite the heavy exploitation, coastal states throughout the region expect increased catches, both from harvesting those stocks that still are lightly fished, and from taking a greater share of the available offshore fisheries consequent on changes in the Law of the Sea. Expansion of the fisheries sector, however, will involve the use of scarce resources, in particular of skilled manpower and capital. The use of these resources, in fishing rather than in other sectors of the economy, will require economic and social justification. Fisheries growth and development pose a range of allocational problems – whether to allot resources to one or other method of fishing, or to domestic or overseas marketing. These choices can be made only through rational economic decisions within a national development plan.

Justification for government involvement in the decision-making processes in fisheries is the open access to the resource. The well-established exploitation of fisheries by individual decision-making units leads to waste of capital and, if unregulated, eventually to economic and social distress.

The problems of availability, exploitation, and management of fish resources in the West African region indicated an important need for additional monitoring and international collaboration. This would ensure improved management, conservation, and shared use of the fisheries resources of the region through the periodic assessment of current stocks and the determination of potential yields. FAO recognized this need and in 1967 established the Committee for the Eastern Central Atlantic Fisheries (CECAF). Membership now includes 21 African nations along the coast and a number of non-African nations that fish the area. At its second session in 1971, CECAF requested FAO to formulate, in consultation with its members and UNDP, a proposal for UNDP assistance for the development of fisheries in the Eastern Central Atlantic. Preparatory assistance for this purpose was approved by UNDP in July 1973 and implemented in January 1975.

Objectives

The long-range objectives of the project were to promote cooperation among the members of CECAF in the management and development of fishery resources within the CECAF area and, as a priority, to increase the ability of coastal states to participate in these fisheries. The immediate objectives were to:

- Improve fisheries statistics and other data;
- Develop a system for continued monitoring of resources and evaluation of stocks;
- Enhance the ability of participating countries to manage the resources and coordinate development planning;
- Assist in the development of national programs and projects that have

significance for regional and subregional development, including aquaculture;
- Train personnel able to implement these activities.

The rationale

The UNDP contribution consisted mainly of expert services, administrative support personnel, subcontracts for resource surveys, and in-service training. The committee sought additional external assistance from non-UNDP sources for the development of associated projects in the overall work program.

As their counterpart contributions, participating governments provided local services and facilities, including buildings for the project headquarters and funded participation by national personnel in research activities and meetings.

The rate of development of the fishing industry in this area depended on the availability of trained manpower throughout the industry, including administrators, research workers, industrial operatives, fishermen, and managers. The project therefore would engage in designing, organizing, and assisting with particularly needed short-term training for all coastal countries.

The assistance that CECAF requested from Canada was aimed at the transfer of the technology needed to strengthen fisheries management at the regional, subregional, and national levels. This assistance was in fields where Canada had recognized international competence – fishery management, fish technology, and utilization – through training seminars, workshops, and courses. Concurrently, identification sheets to illustrate the main commercially important fish species caught, would be produced (bilingually with Canadian assistance) to aid in fishery statistics, biological sampling, resource surveys, and market promotion.

The project proposal was thus divided into two sections covering the management of fisheries and the use of fish technology. After discussions with government fisheries staff in the coastal countries, specific activities were identified and designed to meet the priority management and development needs of East Central Atlantic fisheries.

- In fishery management: Through a seminar and workshops, coastal countries of CECAF would benefit from analysis and review of legal, biological, economic, and social aspects of fishery management. In particular, the activities would centre on practical problems in making optimum use of fishery resources to fulfil national objectives, *viz*, to optimize fish supplies, to maximize economic returns, and to increase employment in the fishery sector.

- In the use of fish technology: Workshops, seminars, and training courses would provide practical guidance to participants from the coastal countries in modern hygienic practices of fish handling and processing, to improve use and quality of marketed products. The majority of the CECAF countries relied on artisanal and near-coastal fisheries as their main source of fish supply. Better methods of fish handling, marketing and processing would help to overcome difficulties encountered at the numerous exposed fish-landing sites in the area. This practical form of development would also help reduce countries' reliance on expatriate involvement and would bring economic benefits to numerous local operatives engaged in the fishing industry. The high levels of wastage and spoilage of fish and fish products in the area needed to be reduced for efficient expanded production and use in the domestic market.

Implementation

FAO, through the program leader of the UNDP/FAO CECAF project, was reponsible for the execution of the project, which was implemented under arrangements between CIDA and FAO. While overall responsibility for implementation rested with FAO, the detailed planning and direction of subprojects was the responsibility of both Canadian and FAO staff who worked closely with CECAF project staff and the national staff in host countries. It was specifically proposed that the director of each training activity would be a national from the host country. He would be responsible for local administration and would be assisted by co-directors from Canada and FAO who would be responsible for the more technical aspects of the activity. Several experienced consultants were recruited with the assistance of the Canadian Departments of Environment and Fisheries and Oceans.

Inputs

The major thrusts of the CECAF program have been to organize the system of fish-capture statistics, their collection and analysis for effective management of the resources, and training of technical officers of the coastal countries. These activities involved a review of the various

Fig 26 Artisanal fishery infrastructures present a particular challenge

fishing systems, artisanal and industrial, from all coastal and foreign fleets fishing the resource, as well as examining the available stocks of important species.

The range of the program activities included specific studies on:

- *Fish resources:* stock evaluation; standardization of methods; sampling methods; statistics collection and analysis; age-reading; tagging; acoustic and trawling surveys for coastal, demersal, and pelagic stocks; species of particular significance (sardines, sardinella, cephalopods, hake and seabream, shrimp, juvenile tunas); localized zones and coastal lagoons; international surveys by research vessels from countries (such as France, Norway, Spain and the USSR) with African observers aboard, while the technical data is analysed and published by FAO.
- *Fisheries exploitation:* artisanal fisheries in Benin, Ghana, Nigeria, the Gambia; rural fish handling facilities; fishermen's cooperatives; small-scale fisheries gear; motorization; marketing and processing; operations of industrial vessels; industrial investment; Las Palmas-based international trawler fleet; joint ventures; economic benefits; plant sanitation; quality control; fish prices and trade.
- *Management measures:* consultations on analysis of data on specific stocks in different zones; mesh size regulations; mesh measuring techniques; implementation of management measures; jurisdiction and enforcement of fishery regulations; monitoring; control and surveil-

lance; draft comprehensive legislation for Liberia, Sierra Leone; legal and institutional aspects of fisheries in the region; compendium of all fisheries legislation of all CECAF countries.

- *Development planning:* model for planning fishery development in the CECAF region; national fishery development plans; alternative strategies for development of marine fisheries; fishery development acts; arrangements for subregional cooperation; individual reviews of fisheries status in Togo, Guinea, Sierra Leone, Liberia, and Cape Verde.
- *Fisheries intelligence and information:* comprehensive fisheries bibliography; newsletters; statistical bulletins; progress reports; marketing information; fishery education and training review; periodic CECAF progress overviews.

Inputs contributed by CIDA within the CECAF program included:

- Seminar on the changing Law of the Sea and the fisheries of West Africa (1978).
- Workshop on fishery development planning and management (1980).
- Seminar on fishery resource evaluation (1978).
- Regional training course on fish handling, plant sanitation, quality control, and fish inspection (1979).
- Regional seminar of senior fish-processing technologists (1978).
- Publication of lectures on fish resources evaluation (1978).
- Publication of species identification sheets for fishery purposes (1977–1982).

The cost to CIDA for the workshops and seminars totalled $1 million. Additionally, the publication in English and French of all documents and fish species identification sheets was directly financed by CIDA. The workshops and seminars were conducted through the program staff with senior specialist technical personnel provided in collaboration with Canadian fisheries institutions and agencies.

For each sub-project published documents have comprehensively described the lectures, technical demonstrations, local or area studies and field activities involved. The workshops were attended by people from almost all the countries within the region. Case studies focused on local problems, with case histories of similar activities in other areas analysed and reviewed. Selected lectures on marine fish resources evaluation, specially prepared for the courses, have been issued, like

Fig 27 Sanitary fishery products, fish preservation and quality control are advocated

other documents mentioned, as FAO/CIDA/CECAF publications for use by the fisheries personnel of the coastal countries.

Outputs

For the CIDA component of the CECAF program, 102 trainees from 20 coastal member-countries of CECAF participated in the various workshops, seminars, and training courses. Manuals and instructional documents were provided for these courses. Ten publications were issued of these activities, and seven volumes, each in English and French, were published of the fish-species identification sheets.

Additionally, in the general activities of the program, four standing committees were established as working parties, comprising scientists from CECAF countries. They have examined available data on demersal stocks, regulatory measures, sardinella stocks, statistics and management, artisanal fisheries, surveillance, etc. During 1975–82 there have been 32 bilingual meetings reviewing resource assessment and management data and 13 training meetings. Twenty-seven documents, 14 newsletters, and 143 technical reports have been issued in English and French, on a range of subjects pertaining to fisheries development and management of the area.

As this region comprises coastal countries in which official languages are English, French, Spanish, and Portuguese, intercommunication in fisheries had been difficult. It was a significant contribution to have documents that precisely describe standard measures agreed on for the region provided to all government fishery offices in English, French, and some in Spanish. Having the technical fisheries personnel working in collaboration had valuable results in coordination of management policies and measures, such as the introduction of uniform mesh sizes for all international trawlers fishing the resource. Without such a program it would have been impossible for the many small countries, with their nascent fisheries services, to have any significant input opportunity for management of the shared fishery resources.

The collaboration by all countries in CECAF has resulted in initial efforts for effective management of the multispecies stocks. Countries have come to appreciate that accurate catch statistics are a basic requirement for fisheries management. As well, policy-makers and investors in industrial fishing companies have become aware of the need for fisheries regulations. Accordingly, the potential and problems of the regional resource are better recognized, and appropriate measures are being adopted. These activities will need to continue beyond the project's termination date.

The bilateral provision under UNDP/FAO auspices of the services and documentation described, by a country not actively fishing the resource, served objectively to guide the technical personnel of coastal countries in their negotiations with major fishing nations. It also was the fore-runner of other bilateral inputs under CECAF.

Effects and impact

The CIDA component of the project catalysed action on resource management, fish processing, and product standards among African coastal countries. As an integral part of the general CECAF program it contributed to greater awareness of management and improved services in some of the following areas:

National

- *Administration and management:* Statistical services; data as a strategic tool; policy planning; development programming; upgrading of fishery services; collaboration with legal institutions for EEZ.

- *Legislation and regulation:* New laws harmonized with neighbouring countries; fishing regulations; application of EEZ; monitoring, control, and surveillance.
- *Technical services:* Artisanal fisheries infrastructure, extension, and training; roles for private enterprise in production, marketing, and processing; improvement of processed fish products.
- *Industrial activities:* Rising national share in increased production; advice on joint ventures; rationalized assessment of investment benefits; performance of national fleets or state fisheries corporations; role of foreign-flag industrial fleets.
- *Fisheries personnel:* Increased theoretical and practical knowledge in specific area problems with unified approach to solutions; sharing of relevant experience with coastal countries where no previous information exchange existed; joint training.
- *General:* A better appreciation of national priorities in regard to fisheries investments; role of artisanal sector; opportunities for national and international research; cost and value of monitoring, control and surveillance.

International

- *Regional collaboration* in management measures, implemented for data collection and analysis and subsequent regulations; harmonization of legislation on fishing zones, foreign fishing operations, and the EEZ.
- *Regional fishing enterprises,* developed with customary and new training partners between African coastal countries.
- *International fishing enterprises,* established by coastal countries with new countries (from outside the region in Eastern and West Europe, Kuwait, Japan, and Korea) on more equitable conditions after review of joint-venture draft agreements.
- *Technical assistance* coordinated through the UNDP, providing for resource surveys, special studies, and capital investments through grants-in-aid, as well as other training programs and workshops.
- *Increased overall production of* the CECAF area with an increased proportion of national landings and heightened awareness of need for effective monitoring and surveillance.

Current situation

Members of CECAF now realize the problems of management of these resources while recognizing the opportunities to get effective results through national, regional, and international actions. Joint programs for training and institutional collaboration have also been proposed. A clearer picture of the complex situation of available resources, (patterns of migration, recruitment, production needs, and future demand) has been emerging. The total production of the region now is about three million tonnes, and only a few northern CECAF countries can be nationally self-sufficient in fish with the resources of their EEZ. The opportunity of sharing the resource through intra-regional trade and collaboration exists. However, the fact that some coastal countries have traditionally earned foreign exchange from extra-regional markets suggests that a trade-off would be necessary. The value of this regional resource is readily gauged by the cost of extra-regional fish inputs required to feed the population when national supplies decline. There are also roles for bilateral assistance from customary trading partners and joint ventures in large-scale industrial fishing, while improving artisanal fisheries in each coastal country. For instance, in 16 coastal countries there are 34 state-run enterprises. Japan, France, Spain, and Kuwait are involved in joint ventures (see references).

Continuity

Only the resource asssessment and statistical analysis part of the UNDP program can be carried out by the FAO regional office and secretariat of CECAF. A promising start has been made through the sponsorship of technical meetings by the coastal countries and their organizations. Thus ECOMAS and CEAO and the Mano River Union (Guinea, Liberia, Sierra Leone) have sponsored training courses and shown interest in fishing investments. Some sharing of training institutions, surveillance services, data processing institutions, and personnel is also being considered. The standards for fish products of intra-regional trade are being reviewed and possible joint enterprises in fish processing may be undertaken. However, much remains to be done to establish permanent, sub-regional cooperation in technical problems for resource management.

Critical outstanding needs have been identified. These include staff training at all levels, extension programs for artisanal fisheries, and the reduction of production costs by lowering inputs of energy that have been subject to inflation. Senior and mid-management staff are inadequate to satisfy national technical production needs (including gear, motors, craft as well as resource management). Considerable marketing and processing inputs are still required for improved product quality, longer shelf-life, and efficient, low-cost packaging of fish products. Expanding operations in the industrial sector also demand more trained nationals to ensure observance of approved practices.

Evaluation and assessment

The workshops and technical meetings were successful in educating policymakers as to the nature of problems and in improving the technical capability of fisheries personnel. It was timely to have those inputs when investments in increased state fishing enterprises and joint ventures were being considered, and while inflated production and equipment costs coincided with the introduction at the U.N. conference on the Law of the Sea of the EEZ for fishing and other purposes.

The problems are becoming clearer, and there is a better appreciation of the interdependence of CECAF countries in the wise use of their common fisheries resources. However, the urgent need continues for sub-regional cooperation, through established institutions, to ensure effective information exchange, multinational management and equitable sharing of these resources. The intervention role of government, its function in management services, and in encouraging enterprise have been well described. It is necessary to ensure that optimum levels of effort are applied; there is evidence that many countries had overinvested in fishing capacity. The introduction of fishery controls and careful study of the impact of these and other measures should be continued throughout the region. There are still wide gaps in the information base that have to be progressively reduced by the collection of more adequate data. This is an area of considerable economic, political and social complexity. The CECAF approach of dividing the fishery into ecologic zones and species groups, while helping to clarify national priorities, has been useful. Nevertheless, it requires the continued application of the best available scientific advice if lead institutions in the CECAF area are in future to assess and interpret the most accurate data that can be collected, so that benefits can continuously accrue to the fisheries as a whole.

For efficient results the continuous monitoring of fishing operations, investments, and operational costs will require better, reliable, and continual communication between fishermen, administrators, and scientific assessors. This is perhaps already happening between governments and industrial users. In this regard the various workshops and publications have highlighted the conflicting interests in the large CECAF fishery, internationally as well as nationally. Whereas the international problems remain formidable there are still many such opposing interests, within national jurisdiction, that are not yet resolved – those of commercial fish traders versus cooperative operations or coastal versus mechanized fishing systems, for instance. Whereas CECAF therefore has provided an international forum for discussing benefits and constraints, similar interchanges of ideas for determining wise national priorities and avoiding surprising local failures of socially motivated developments are necessary.

The overall impact has therefore been evident in the increased fisheries management activities, nationally and regionally. The greater interest in fisheries, however, takes a long time to be transformed into the returns on investment that governments and development banks seek. Perhaps this is because the limited impact of state fishing operations has discouraged precipitate investment in more large national enterprises. However, artisanal fisheries and joint ventures have gained increasing prominence in total fish production.

We should not neglect the continued need for training of national staff so as to establish at recognized lead institutions, a permanent regional cadre of technical and policy administrative expertise. Further follow-up refresher courses by bilateral donors may be timely. The total value of this regional resource is more readily gauged by the cost of extra-regional fish imports that are required to feed populations when national supplies decline. Accordingly governments should appreciate the value of the needed investment for upgrading skills.

It may be useful to compare these developments and activities with those undertaken by a somewhat similar inter-regional program, the South China Sea Program. The regional institutions or lead centres and the use of information for investment in that area are a worthwhile subject of study.

The seminars that were conducted by CIDA for senior management and fisheries decision-makers, as well as the workshops on processing and quality control were useful. It is evident that updated and specifically focused seminars of this nature from bilateral donors are needed inputs to such programs. The need for a regional training course on monitoring, control, and surveillance of the fisheries in the EEZ has created considerable interest among West African coastal countries. This is a field in which Canada has considerable relevant experience and increasing expertise.

Selected references

Ansa-Emmim, M. 1979a. Availability of biological data (mainly length/frequency) for stock assessment in the CECAF countries. In report of the fourth session of CECAF Working Party on Resource Evaluation, Dakar, 23–27 April. Rome, Italy, FAO, FID/R220. 161–179.

1979b. Mesh sizes used for trawl fisheries in the CECAF area. In report of the fourth session of CECAF Working Party on Resource Evaluation, Dakar, 23–27 April. Rome, Italy, FAO, FID/R220. 187–197.

Ansa-Emmim, M., Mizuishi, I. 1980. A summary overview of fisheries in the CECAF region. Rome, Italy, FAO, CECAF Technical Report 21/E. 62 p.

Ansa-Emmim, M., Levi, D. 1975. Biostatistical data for stock assessment purposes: present situation and suggestions for improvement. Rome, Italy, FAO, CECAF/ECAF/75/2. 16 p.

Carroz, J., Moore, G. F. K. 1978. Possible institutional arrangements for subregional cooperation in fisheries among the governments of Cape Verde, Guinea, Bissau, Mauritania, Senegal and The Gambia. Rome, Italy, FAO, CECAF Project Report (also in French and Portuguese). 11 p.

Christy, F. T. 1979. Economic benefits and arrangements with foreign fishing countries in the northern sub-region of CECAF: a preliminary assessment. Rome, Italy, FAO, CECAF/ECAF/79/19. 39 p.

Evans, E. D. 1981. Report on mission to draft comprehensive fisheries legislation for Sierra Leone. Rome, Italy, FAO, FAO–CECAF project GCP/RAF/146 (NOR). 60 p.

1982. Report on mission to draft comprehensive fisheries legislation for Liberia. Rome, Italy, FAO, FAO-CECAF project GCP/RAF/146 (NOR). 14 p.

Everett, G. V. 1976. An overview of the state of fishery development planning in the CECAF region. Rome, Italy, FAO, CECAF/ECAF/76/4. 67 p.

1978. The Northwest African fishery: problems of management and development. Rome, Italy, FAO, CECAF Technical Report 6/E. 43 p.

1979. Summary of current fishery development plans. For the Sixth Session of CECAF, 1979, Agadir, Morocco. Rome, Italy, FAO, CECAF/VI/79/Inf.13. 9 p.

1981. Progress of CECAF Project activities. For the Seventh Session of CECAF, Lagos, Nigeria, 1981. Rome, Italy, FAO, CECAF/FD/IV/81/Inf.6. 12 p.

FAO (Food and Agriculture Organization). 1975. A bibliography of West African marine fisheries and fishery oceanography. Rome, Italy, FAO, CECAF/ECAF/75/1. 145 p.

1976a. A bibliography of West African marine fisheries and fishery oceanography. Rome, Italy, FAO, CECAF/ECAF/75/1 suppl. 1. 73 p.

1976b. Report of the FAO/NORAD mission to West Africa on the NORAD project proposal: Pilot project for handling, processing and marketing of fish. Rome, Italy, FAO, W/K 1657. 30 p.

1978a. CIDA/FAO/CECAF regional training course in fish handling, plant sanitation, quality control and fish inspection. Dakar, Senegal, 10 October – 4 November 1977. Rome, Italy, FAO, FAO/TF/INT 180(f) (CAN). 52 p.
1978b. Report of the *ad hoc* working group on coastal demersal fish stocks from Mauritania to Liberia. Rome, Italy, FAO, CECAF/ECAF Series 8/E. 8/F. 98 p.
1978c. Report of the CIDA/FAO/CECAF regional seminar of senior fish processing technologists, Dakar, Senegal, 10–14 October, 1977. Rome, Italy, FAO, FAO/TF/INT 180(g) (CAN). 30 p.
1978d. Report of the CIDA/FAO/CECAF seminar on fishery resource evaluation, Casablanca, Morocco, 6–24 March. Rome, Italy, FAO, FAO/TF/INT 180(c) (CAN). 14 p.
1978e. Report of the working group on standardization of age determination of the sardine (*Sardina pilchardus* Walb.). Instituto Espanol de Oceanografia, Santa Cruz de Tenerife, Canary Islands, Spain, 17–19 April. Rome, Italy, FAO, CECAF Technical Report 8/E. 8 p.
1978f. Report on the CIDA/FAO/CECAF seminar on the changing Law of the Sea and the fisheries of West Africa, Banjul, The Gambia, 19–27 September, 1977. Rome, Italy, FAO, FAO/TF/INT 180(a) (CAN). 147 p.
1978g. Report on the investment feasibility proposals for development of the sardinella fishery in the Mano River Union (January – February 1978). Rome, Italy, FAO, TF/RAF 80 (j)(NOR). 15 p.
1979a. Catalogue des engins de pêche artisanale du Sénégal. Rome, Italy, FAO, CECAF/ECAF/79/16. 110 p.
1979b. Report of the *ad hoc* working group on fishery planning. Rome, Italy, FAO, CECAF Technical Report 79/14. 63 p.
1979c. Report of the special *ad hoc* working group on the evaluation of demersal stocks of the Ivory Coast – Zaire sector. Rome, Italy, FAO, CECAF/ECAF/79/14. 74 p.
1979d. Report on the consultation on stock management in the CECAF statistical divisions Sahara and Cape Verde, Dakar, Senegal, 18–22 June. Rome, Italy, FAO/SCP/RAF/148 (NOR). 24 p.
1980a. Selected lectures from the CIDA/FAO/CECAF seminar on resource evaluation. Casablanca, 6–24 March, 1978. Rome, Italy, FAO, FAO/TF/INT 180(c) (CAN) Suppl. 166 p.
1980b. Rapport de la réunion spéciale sur la mesure de l'effort de pêche appliquée aux petites espèces pélagiques dans la zone nord du COPACE. Rome, Italy, FAO, CECAF Technical Report 80/19. 64 p.
1980c. Report of the CIDA/FAO/CECAF workshop on fishery development, planning and development, Rome, 6–17 February, 1978. Rome, Italy, FAO, FAO/TF/INT 180(b). 484 p.
1980d. Report of the preparatory study for establishment of a regional fishery cooperative development centre in the CECAF region. Rome, Italy, FAO, GCP/RAF/804 (NOR). 28 p.
1981a. A manual on acoustic surveys: sampling methods for acoustic surveys. Rome, Italy, FAO, CECAF/ECAF/80/17. 47 p.
1981b. Report of the consultation on management of cephalopod stocks in CECAF statistical divisions Sahara and Cape Verde Coastal. In Report of the

Third Session of the CECAF Sub-Committee on Management of Resources within the Limits of National Jurisdiction, Dakar, 28–30 January. Rome, Italy, FAO, FID/R250/E. 18–44.

1982. Report of the consultation on artisanal fisheries in the CECAF region, Dakar, Senegal, 1–4 September, 1981. Rome, Italy, FAO, CECAF/TECH/82/39. 29 p.

Fidell, E. R. 1978a. Legal and institutional aspects of fisheries management in the Republic of Liberia. Rome, Italy, FAO, CECAF/TECH/78/11. 38 p.

1978b. Legal and institutional aspects of fisheries management in the Republic of Sierra Leone. Rome, Italy, FAO, CECAF/TECH/78/12. 39 p plus annex.

Griffen, W. L., Grant, W. E., Shotton, R. 1982. A bio-economic analysis of a CECAF shrimp fishery, and operations and background data. Rome, Italy, FAO, CECAF/TECH/82/41. 78 p.

Gulland, J. A. 1979. Towards the management of the resources of the CECAF region. Rome, Italy, FAO, CECAF/ECAF/79/12. 19 p.

Hamlisch, R., Moore, G. F. K. 1975. Joint ventures in fishery development in the CECAF area. Rome, Italy, FAO, CECAF/ECAF/75/3. 25 p.

Jennings, M. G. 1980. The enforcement of fishery regulations. Rome, Italy, FAO, CECAF/TECH/80/22. 23 p.

Laming, G. N., Hotta, M. 1979. Fisheries cooperatives in West Africa. Rome, Italy, FAO, CECAF/TECH/79/17. 18 p.

Pereira, J. A., Bravo de Laguna, J. 1981. Dinámica de la poblacion y evaluación de los recursos del pulpo del Atlántico Centro-oriental. Rome, Italy, FAO, CECAF/ECAF/81/18. 53 p.

Purcell, A., Seck, B., Everett, G. V. 1978. Report of the CECAF project mission on the possibility of setting up a fish marketing information and technical advisory service in West Africa. Rome, Italy, FAO, CECAF Project Report. 27 p.

Robertson, I. J. B. 1977. Summary report: Fiolent 1976. Eastern Central Atlantic coastal fishery resource survey, southern sector. Rome, Italy, FAO, CECAF/TECH/77/2. 115 p.

Ruppin, R., Deltour, J. P. 1977. Working paper on fishery education, training and extension in the CECAF region. Rome, Italy, FAO, CECAF Project Report. 21 p.

Strømme, T., Saetersal, G., Gjøsaeter. 1982. Preliminary report on surveys with the R/V "Dr. Fridtjof Nansen" in West African waters, 1981. Presented to the CECAF Working Party on Resources Evaluation, 2–6 February. Rome, Italy, FAO, GLO/79/011. 69 p plus annexes.

Talarczak, K., Haliog, L. 1978. Planning of fish handling facilities for a rural fishery centre. Rome, Italy, FAO, CECAF/TECH/78/9. 19 p.

Talarczak, K., Mizuishi, I. 1977. Industrial marine fisheries in the CECAF area. Part II: Ivory Coast to Zaire. Rome, Italy, FAO, CECAF/TECH/77/5. 71 p.

10
Conclusions and recommendations

Historical review

Initial projects

The foregoing analyses of the operations of fisheries development projects in many parts of the world reveal a number of common features encountered during the past 15 years, to a greater or lesser extent, in carrying them out. It was a period when there was major international funding support for national and regional fisheries projects. Thus many countries and agencies were gaining experience, and weak fishery departments were also 'learning by doing'. Programs started hesitantly and passed through changes as staff training, services, and fishing installations became better established; donor agency procedures meanwhile became more applicable to the particular circumstances of fisheries development.

Most of the projects analysed constituted part of a general overall development plan of a government to augment food supplies and improve employment opportunities in fisheries. Therefore, the projects involved aspects such as resource assessment and management, harbour facilities, construction and maintenance facilities, production systems and services, marketing and distribution amenities, as well as training institutions. During the formulation of the projects top policy decisions were required on the level of the priority attached to these activities by international and regional organizations, as well as by central governments, or local authorities for the administration and implementation of these programs. Whereas strategies and directives for fisheries development were broadly agreed on, the actual planning, design, and implementation of projects often revealed constraints which were inconsistent with policy objectives.

This chapter seeks to highlight the lessons learnt through these analyses, which one hopes will be useful to fisheries planners and administrators in many countries. Within the historical context of the past 15 years, national priorities will be reviewed, regional programs compared, project planning and implementation assessed, operational constraints identified, comments on loan policies offered, and, finally, guidelines recommended for successful implementation of projects.

Knowledge of fisheries resources

Much information was being gathered about the occurrence, availability, and variation of fish stocks. However, on the whole, less attention was being paid to the specific performance of species in the areas where fishery development projects were being implemented. Greater success was evidently experienced either where resources were abundant or, where they were less so, there was ample knowledge of the seasonal or migratory patterns of the stocks. When preparatory data were not enough to provide reliable information on fishable stocks, ventures in fishing operations and investment were less secure. The improved organization and collection of statistics and the monitoring of catch data proved to be important strategic tools for management. As projects developed, such resource data helped provide guidelines for investment. Projects that began with inadequate data-bases encountered surprises. Acquiring knowledge of tropical multi-species fish stocks and their behaviour was particularly difficult. Exploratory surveys and data collection were most successful when initially confined to a limited area and later expanded. Such data had to be collected over several seasons before interpretations provided valid indications for the fishery. Accordingly, as projects proceeded, the data provided more information as to the cyclic occurrence and behaviour of fish stocks for effective management.

External factors

During the period there was world-wide inflation in commodities. The most important of these was fuel, which in some areas quadrupled in price. This had a general deterrent effect on motorization and on fuel-intensive fishing systems. However, the cost of vessels, outboard engines, fishing gear, as well as storage and processing equipment and services also increased. Governments could best assist by providing all equipment for the industry with a minimum of tax added to their import cost so as to lessen the effect of inflation.

The new Law of the Sea introduced the exclusive economic zone (EEZ) for fisheries. This provided a new opportunity for developing countries but increased their responsibility for management of the fishery resources adjacent to them. This occurred concurrently with their efforts to organize their staff, infrastructures, and services for fisheries development.

Major lessons

The preparation and implementation of the projects clearly provided considerable new experience for all administrations involved. It soon became evident that there is no such thing as an instant fishing industrial development, and the complexity of problems encountered soon made government and administrations appreciate much better the difficulties and commitment needed of a long-term effort. It became clearer that the coordinated efforts of the administrative services, the private sector, and the institutional training and extension services must all contribute to fisheries development.

National priorities

Commitment and policies

Effective fishery programs, which may be collections of related projects, must cover the multidisciplinary aspects of fisheries. They must reflect the resolve and commitment of governments to them as instruments of policy. Various aspects of such a positive long-range program will affect fishermen, processors, marketers, and consumers. Consultation (between the private sector and the fishery administration in the formulation and execution of projects) can enable consistent supportive policies to be established and operations to be modified when needed. Such policies permitted equipment for fish harvesting, processing, and distribution to be made available at low cost. Incentives in some cases favoured emphasis on local food supplies and employment, with balanced promotion of fish production to earn foreign exchange.

Coordinated activities

Successful project activities were generally integrated with community development programs toward national perspectives. Projects that were confined to district activities or remained specific departmental operations suffered from lack of coordination and had limited impact. Projects whose activities incorporated district, national, and inter-country fishing systems stimulated more attention from governments and the private sector. The consultation and review process between government administrations and fishing operatives when introducing changed operational systems, was essential for overall success, because it helped to reinforce the initial commitment of government, as well as to provide guidelines for planners as the beneficiaries reacted to project activities.

Training

Project organization and investment procedures were considered by most project countries to be too long. There is an evident need for training national staff in project financing procedures within national agencies and for the requirements of overseas organizations lending systems. Regional groupings of countries could undertake this training. Most fishery departments were headed by biologists unaccustomed to furnishing the data or following the procedures that investment agencies require and readily obtain in other natural resource industries. Those concerned with policy planning and development also tended to lack this knowledge of international funding procedures. Enlightenment is needed at this level also.

Fishery education and training has been declared a priority by most governments during the past three decades. However, technical staff to implement these types of projects required training on-the-job or from extension services, which had in many cases to be organized during the projects. Investment in training of fishing operatives is an important long-term commitment that cannot be abandoned half-way or neglected without serious adverse effects.

Artisanal versus industrial fisheries development

Throughout the developing regions the conflict between artisanal, traditional, inshore fishing and industrialized fishing has been apparent. The artisanal fisheries have been clearly recognized by governments as an important source of food and employment. Their social and political significance has been clearly accepted, but the need remains to measure the harm that traditional methods do to available resources, and to introduce measures for their progressive elimination while improving efficiency and ecological stability of the fishery. Artisanal fisheries development presents a dilemma and challenge; it involves:

The improvement of fishing craft; motorization; efficiency of operations; limitations of investment; assurance of adequate financial returns; constant evaluation of practices to ensure profitability, improve processing and storage, provide stable products, and avoid wasteful practices through diversification; improved community organization; and comprehensive planning.

With the development of extended national jurisdictions and eventual promulgation of the EEZ, the allocation and management of resources being exploited by artisanal fisheries becomes more significant. Furthermore, the contribution of industrial fishing in offshore waters to national fishing needs should be carefully reviewed. Thus, the strategic modernization of the fishing industry will require a long-term view and a comprehensive assessment of the contribution that the artisanal sector can really offer, while providing improved living conditions. These have clearly to be developed with local guidelines to ensure profitability, improved health and earnings, and social uplift of the fishing communities.

Role of aquaculture

Another constant theme in countries whose fisheries are less developed has been aquaculture. Inland and coastal fish husbandry has played a significant role in southeast Asia, where aquaculture is most developed worldwide. However, this still represented less than 10% of fish production and less than 5% of the fisheries development budget of most southeast Asian countries. The significance of inland fisheries when planning maximum production is shown in chapter seven. However, as marine fisheries are already exploited close to the limits of their potential in many countries, there will be an increasing emphasis on aquaculture to enhance food production in inland and coastal waters.

Funding policy and credit

Credit conditions, the assessment of risk, loan criteria, revolving funds, and other aspects of investment in fishing should be the subject of special regional concern by governments and investment agencies, as well as of specific review within countries. Related to this are the limits that have to be introduced to avoid over-investment such, for example, as too many vessels fishing a resource, too many outboard-powered inshore craft, or too many processing installations, cold-storage warehouses, boatyards, or slipways. Whereas in one case the fuel crisis curtailed private investment in further boatyards and iceplants, the establishment in other areas of public or private facilities resulted in increased competition, in decreased profitability for some fishing operations, and even, eventually, in their decline. Accordingly, the control of credit for certain fishing systems could be a useful management tool to limit investment and provide for strategic modernization, particularly in the artisanal sector.

Incentives have been important in stimulating and sustaining new operations. These have notably included tax-free fuel for artisanal and industrial fishing vessels, tax holidays for new fishery installations, low-interest loans, insurance for inshore craft and outboard engines, market guarantees for fishermen, and minimum prices for catches. These incentives can also encourage overcapacity in the industry and should therefore be carefully and sparingly used. Such measures to encourage new enterprise were seen to be part of a national policy to encourage employment and self-sufficiency in food with consequent gains of foreign exchange through import substitution and export earnings. In many cases these incentives were a time-limited subsidy to the industry; industrially developed fishing nations had also subsidized their fishing industries, some of which operated off the shores of developing countries, notably in the East Central Atlantic and the South China Sea. It is important that developing fishing enterprise should not compete, using taxed and costly imported equipment, against imports of fish produced by subsidized industries of developed countries.

Local interests and regional collaboration

Because of the common-property nature of shared fishery resources, local development cannot ignore the need for collaboration with adjacent countries and regional management of the resource. Harmonized regulations and joint use of regional institutions and services offer a good opportunity for effective management and savings.

Regional programs

Resource management

Both regional programs reviewed here have shown that joint collection of catch data holds clear advantages for the efficient management of fishery resources harvested by several nations. The systems of data collection, analysis, stock assessment and the introduction of regulations, have been well illustrated in the programs of CECAF and SCSP.

Additionally, the technical need to introduce regulations has been amply justified to countries of the region.

Training and research institutions
Whereas on-the-job training has been a part of every project, much regional and international experience in designing suitable training systems for particular situations is available. It is clear that the establishment and operation of first-class fishery schools or research institutions are expensive for an individual country and require adequate staff, which is often unavailable, at least initially, within the country. In such cases a regional approach might be possible although there are many problems, such as political, economic, cultural, and linguistic differences. However, if each country within a region were to specialize in disciplines of particular interest to it, this would permit a sharing of investment in instructional services. The development of research lead centres can be equally valuable where certain activities require heavy investments in human resources and equipment. Activities for which such a collaborative approach was used in SCSP included stock assessment and analysis, coastal zone and mangrove ecological studies, pesticide toxicology and pollution, product development for underused species, and breeding techniques of important fish and invertebrates in aquaculture. Again, with regard to the Law of the Sea, special attention could be paid to the need for regional approaches to, and surveillance in, the adjacent EEZ of countries.

Standardization
The regulations and technological developments introduced in the fisheries of the South China Seas and East Central Atlantic described in chapters eight and nine reflect the standards applied and the impact they are having on the exploitation of the total resource shared by those countries. Similarly, the approaches that SCSP and, in some cases CECAF, have established for standardizing procedures in joint-venture negotiations, pollution evaluation, and small-scale fisheries development would clearly be of benefit to other regions of the world where there is a competitive scramble for resources among countries. Further, the preparation and use of manuals that pertain to production systems or operations could be done regionally, using standard techniques, applied in each country and recommended as the preferred method. This would also facilitate the application of management measures.

Information exchange
Apart from joint preparation of manuals, detailed knowledge of experience with enterprises undertaken in neighbouring countries would be of much value to countries planning similar new enterprises. The sharing of experience has been well developed through regional programs and workshops by SCSP and CECAF, and these may well be continued. Preparation of adequate extension manuals and improvement in systems of information exchange are useful coordinating roles for international fisheries commissions to undertake. Thus they can help standardize the quality of manuals and technical leaflets produced nationally for extension purposes, as well as strengthen national capabilities in regional management of fisheries resources.

Role for bilateral aid
In these programs there has evolved a significant role for bilateral aid as inputs into UNDP regional programs for fisheries management and development. Donor organizations can make distinctive but collaborative contributions to regional programs and to national projects that have regional application. This will further stimulate the effective harvesting, world-wide mobilization, and strategic modernization of underdeveloped fisheries resources in Third World countries. The impact of separate and individual bilateral aid has been shown to be less than that of coordinated programs. Many nationally sponsored programs can share their experience for the mutual benefit of the fisheries services in similar, neighbouring countries. Coordinated projects have been able, in such areas as the North Pacific fisheries, to enhance the total production and profitability of the fisheries and rehabilitate them.

Project planning and implementation

Preparation
The preparation of an internationally financed fisheries program normally would involve missions that first formulate and identify projects and appraise their rationale, and are subsequently concerned with negotiation and approval, implementation, supervision and monitoring, and the terminal evaluation. The planning and scheduling of activities in the preparatory phase of the project require time for indepth studies and the gathering of detailed data, the selection of staff, and the appraisal of the operational realities of each aspect of the project. The administrative

services of countries need to be well geared with data to take advantage of available opportunities; they need to know what data are required to analyse the feasibility of a project and justify it. Apart from this, preparations need to be made to recruit adequate technical support staff before the start of the project rather than during it. Pre-project staff recruitment and training was more effective, as such nationals who received further experience on the job were of service throughout the project.

Training must be a continuous process before, throughout, and after the project's scheduled activities. In most cases the best returns on investment in fisheries development were realized where there was staff continuity for the duration of the project, combined with the gradual enlarging of the infrastructure services. Recruitment criteria should include relevant experience not only in government administration, but in the private sector as well. Other essential preparatory activities are the accumulation of adequate data on resource, operational seasons, cyclic availability, and the markets and shore facilities on which the project outputs will rely. Often there was no continual gathering of data for comparative assessments, monitoring, and evaluation, which are always critical to effective management. Difficulties were encountered where there were no previous data for comparisons.

Installations

Installations for the technical operations of a project were often out of phase with other activities. Delays in site selection, preparation, delivery of essential components, gear, or vessels and other procurement hindrances presented serious difficulties, apart from the shortage of competent staff. These problems can throw the implementation schedule out of phase. Some problems can clearly be avoided by careful planning; others are aggravated by unseasonable weather or unforeseen economic circumstances at the national and regional levels. Adequate preparation and some built-in flexibility are necessary for efficient project operations.

Institutions and organizations

Fisheries administrative structures were often weak in development and organizational needs compared to the forceful private sector. Many project activities required resourceful personnel, often with skills very different from those of the biologist-type people generally recruited. Too often, detailed plans for design and implementation of projects involved unrealistic schedules and goals. Though taking time to be well established the institutional improvements introduced by foreign project management organizations generally contributed significant developments to the fisheries infrastructure services that would otherwise not have resulted despite national awareness of the need. Project advisory or coordinating committees and other special subcommittees are essential supportive bodies. Their function is to review project activities and provide ongoing monitoring and evaluation of results. These reviews should be considered at both the central and district level. Feedback from participants at many of the international seminars and workshops confirmed the value of these reviews. Nationally executed projects involving various departments and services, can also find such committees useful. When action was taken in response to objective analysis provided by project evaluations, developments progressed and difficulties were foreseen and avoided. In some cases delays in installations and implementation required altered timing of staff recruitment or their redeployment to other activities. Such actions are quite normal in commercial fisheries but may be unusual and difficult for new fisheries administrations, particularly within a civil service. Fishery development exigencies often require precedents for administrative procedures which are resisted by bureaucracy.

Regional experience

It is essential that similar programs implemented in the region should be carefully examined and compared so that the benefit of the relevant experience gained can be incorporated in the local project and much time and expense thus saved.

Publicity and communications

Publicity was often inadequate or unsatisfactory. News of the official signing of the agreement for a project created expectations that were often frustrated by the long build-up as the project took shape. Where positive results were publicized, it stimulated administrative confidence to maintain performance. Press publicity tended to emphasize the spectacular. Often this accentuated the negative rather than the positive aspects of projects, but progress is not always newsworthy. However, to their credit, there were public relations demonstrations in many projects that had notable achievements to their credit. Technical reports have been written on all the projects and this has resulted in several articles in national journals or papers at technical international conferences. However, there seemed inadequate national publicity of the successes of new undertakings; positive accounts would encourage industry operatives and inspire confidence in the administration. More emphasis was clearly needed on instructional extension leaflets, manuals, and public information for entrepreneurs. The benefits of regionally exchanging information and standardizing techniques has been well demonstrated in SCSP and CECAF.

Operational constraints

Human resources
The availability of adequate technical management and extension staff was generally a serious difficulty for projects in the initial stage, but subsequently staff training permitted the programs to develop more effectively. Several projects had many changes of management personnel. Staff continuity was also a problem for many of the donor agencies.

Infrastructures
Apart from administrative management there were technical matters that took time to prepare. These included design studies for suitable fishing craft, fishing port sites, cold storage installations, laboratory and processing facilities, and specific techniques for particular species. Where these were to be preliminary activities, the funds or time provided were often not scheduled flexibly enough to accommodate the slow achievement of results without hampering other project activities. Other circumstances, such as unexpected weather, inadequate staff services in new operations and slow procurement of equipment, also caused serious delays.

Upset schedules and budgets were further complicated by the worldwide fuel crisis and general inflation, which adversely affected developing countries and had a severe impact on the fisheries sector. In some cases the management of a project was efficient enough to be able to make prompt decisions locally to adjust and curtail plans so as to pursue timely activities and keep the project alive despite external economic problems. However, slow decisions and exchange of information within other projects described, in Africa and Latin America, created great problems for weak infrastructures.

Lack of efficient communications for field activities, and for the referral of problems to management by users, was a major constraint in the canoe motorization project. In most of the bilateral and investment projects there was room for improvement in disseminating information. Particularly in remote districts with promising fisheries potential, the extension or supportive services needed improved communications to ensure the effectiveness of projects. In general, the public relations aspect of these fisheries projects was given little attention. Although there were many unprecedented achievements, little public approval, and thus encouragement for staff efforts, resulted.

Critical issues

Donor agencies tended to consider the schedules rigidly whereas recipient governments evidently expected that donor inputs would automatically continue during extended schedules. Rarely did any project keep to its planned timetable, and many components undertaken were left incomplete at the end of the planned duration. This is a significant failing, and it is clear that donor agencies should adopt more flexible, but realistic phasing. This is clearly an important consideration in view of the many changing circumstances within a fisheries project.

Program development by gradual phases has evidently been adopted as a policy by some investment agencies. This is designed to match the absorptive capacity of fisheries managers, government administrations, and the private sector involved in production and marketing to adopt changes, particularly in artisanal fisheries. A series of short, medium-scale projects may be more effective as a precursor to a larger investment, when infrastructure and services are better established and those involved have gained stature, experience, and confidence nationally and are more capable of handling international projects. However, for international donor agencies small projects are less attractive since they involve virtually the same preparatory and management costs as large programs. Often these smaller projects are relegated to non-governmental organizations and volunteer services which lack the technical back-stopping to support fishery projects.

Projects generate more support when successful; failures become disincentives to all sectors as well as to donor agencies. Some key elements in operational success are realistic goals and thorough preparatory study. Limits on the scope of activity and the operating area, adequate time during the initial phase, and concentration of staff and efforts on initial field activities will ensure the initial aims are accomplished. When this has been in full collaboration with private fishermen, traders and financiers, it generally led to more lasting results.

Other factors in success were phased and effective staff training, coordination with bilateral and multilateral programs in the same and adjacent countries, advisory and review committees to deal with problems before they became unsurmountable, assured markets providing equitable prices, and adequate supplies of fishing equipment and engine parts.

The response by management to consumer and trader actions was also significant in project success. After it was evident that an initial project had been consolidated, the extension of activities to accommodate systems of a different nature permitted logical, timely planning of supplies, replacements, variations in area services, and markets, with less confusion. Extending the operational system of the project on a national scale requires continued collaboration with the private sector. It is essential to ensure that all current and new activities in different localities are carefully checked for profitability. Thus the evaluation and monitoring reviews should be shared by the administration with users. There has to be a continuous interaction through discussions with the fishermen, consumers, and traders; this dialogue, with publicity of the project's progress, can contribute to lasting development.

Loan policies of development banks

Area banks have given more loans for fisheries development collectively, and in some cases, individually, than the World Bank. It is certainly the case in Asia. The Asian Development Bank provided 27 loans in 14 years of operation up to 1980 totalling about $340 million. This, however, only represents about 4% of the actual $8 billion of bank loans approved to that date. Additionally, there have been 21 technical assistance projects, grants, and contributions in rural development projects.

The 1981 report of ADB describes well the role of fisheries in Asia and the South Pacific and outlines the various activities and constraints in fisheries of the Southeast Asia region. It indicates the levels of bank assistance and the emphasis placed on projects in marine fisheries, aquaculture and inland fisheries, fishing ports, and marketing and processing technology. The institutional characteristics and difficulties of bank-assisted projects with regard to subloans, end users, fishery corporations, cooperatives, and government departments, as well as evaluation of operations and benefits have been all detailed in this compilation.

Several issues in fisheries development investment that clearly pose challenges have to be carefully considered and resolved. These include: extended jurisdiction, the availability and efficient management of fisheries resources, the importance of aquaculture and inland fisheries in the overall production of various countries; additionally the availability of appropriate technology, the option of development of small-scale fisheries rather than larger industrial ones, and the major problem of post-harvest technology. Finally the role of local institutions in fisheries organization, training, and research must be considered. There are also problems in procuring appropriate equipment and in the financing by local institutions of supportive activities.

The Asian Development Bank therefore decided that their future emphasis should be to increase the availability of fish and fish products for domestic consumption, while increasing exports or replacing imports in order to earn and save foreign exchange. These activities would in turn emphasize the aim of generating employment and increasing incomes at all levels of the fishing industry. This sector emphasis was further classified according to priorities of marine fisheries, aquaculture and inland fisheries, resource management, education and training, as well as the institutional framework that should be associated with the development of fisheries. Although it records an impressive series of activities, the report concedes that there is very much to do and there is a considerable shortfall between what is required and what is being undertaken. This open recognition and the magnitude of the remaining problems provide a sobering challenge for bank investment in the area.

By taking an analytical look at the Asian region – one of the most populous, and one of the most important as far as fish consumption and development is concerned, and one where bank investment activities have shown an enlightened developmental thrust – the reality of the dilemma facing world-wide fisheries development can be seen. National governments and international agencies are appreciably sensitized to the need in that area. Yet despite this, the increasing demand has not met an adequate, timely, effective response; the gap between fish requirements and supply is evidently widening. This situation illustrates the sobering and perplexing reality for other regions of the world, where recognition of the importance of the role of fisheries has not yet been as pervasive throughout all levels of government, and where the area banks provide an even smaller percentage of their total loans for industrial, social or economic development to fisheries. As only 4% of the investment budget of the bank has gone into fisheries in that part of the Third World that has most aggressively pursued fisheries development, it is cause for reflection on what more must be done to enlighten and educate top policy-makers and motivate them into a realistic, concerted effort to mobilize the ability to harvest rationally the fishery resources of the aquasphere.

Needs for long-term investment are:

1. More regional and national efforts toward investment in training institutions for skilled manpower for all fishing activities. The calibre and availability of trained staff in the region is the major limiting factor in successfully accelerating the development thrust.
2. New or strengthened institutions and greater incentives for operating critical services. A wider perception of constraints in various national sectors in fishery development may arise only after further appraisal.
3. Re-assessment of conventional public attitudes and investment services available at para-statal and private banks for fishery development. The success of aggressive private enterprise, which is notably efficient in many Asian countries, shows the need for prompter response to demands for fishery credit.
4. Loans are mostly channelled through government banks and the subsequent sub-lending to the private entrepreneur may be influenced by considerations other than objective assessment of his capacity. Criteria for providing effective credit should be reviewed and include pragmatic as well as development considerations.
5. The coordination of effort and the implementation of rational fishing practices have to be ensured through continuing stock assessment and collective management of shared resources under government arrangements.
6. In this regard, then, permanent regional training institutions are absolutely vital to the effective, sustained management and harvesting of the resource. Therefore, in institutional support and training there should be a fairly long-term investment in order to ensure that the operatives and producers who will manage development activities within the next five to 10 years are indeed trained in the most enlightened, versatile, and efficient systems, and can cope with their

fisheries in small-scale activities, aquaculture, or marine industrial enterprises.

Role for bilateral aid in multilateral programs

Coordinating committee personnel and government representatives have indicated their general satisfaction with the work accomplished by the programs in improving development perspectives for regional fisheries through coordination and sensitization of governments and industry. Though opportunities for development have been reiterated during the 30 years of deliberations of FAOs Indo-Pacific Fisheries Council, SCSP served to precipitate immediate developments by the private sector and bank investment in some countries. It also created an atmosphere, through workshops, of friendly development rivalry between countries, for improved planning organization and regulatory management measures. During the implementation of the program, the emphasis of governments shifted from the development of large-scale marine fisheries to small-scale artisanal fisheries, education, and extension services; as well there was a greater emphasis on aquaculture. Governments were reorganizing and mobilizing their services in fisheries, while the private sector in different countries became more aggressive and efficient in implementing changed operational systems that were demonstrated or discussed as part of SCSP activities.

Although this is the area of greatest aquacultural activity in the world, aquaculture is of more social and employment importance in rural areas and contributes no more than 10% of total fish production in any of these countries. Conservation of the coastal mangrove areas as a nursery for marine fish was recognized as important mainly through adverse effects of its destruction caused by pollution, mining, or other circumstances. The percentage of fish produced for food from coastal fishponds is small compared to marine production, though the local employment directly generated from coastal ponds is often considerably greater than that involved in sea fish production. The multiplier effect, therefore, of aquaculture is recognized by most governments as socially significant and politically important. Freshwater fish culture has also had an increasing importance in some countries where marine fish harvests have diminished.

The South China Sea Program had succeeded in obtaining and implementing 29 different projects with trust funds (from several countries including Canada, Japan, Australia, Norway, Sweden, the Philippines), apart from the funds provided by the UNDP inter-regional program. For instance, three Canadian projects provided basic support funds but also generated income from fish sales from the pelagic survey. Accordingly, Canadian inputs totalled $5 989 million in 1974–1982.

During implementation, the coordinating committee of SCSP developed a significant management function. Initially it was intended as an advisory and coordinating body with representatives from all participating countries. It subsequently assumed almost executive functions. It reviewed and approved the program proposed by technical management, determining priorities for funding. It evolved into the Committee for Development and Management of Fisheries of the South China Sea. Further activity will continue through the FAO regional office and national or regional bodies, particularly the Association of South-East Asian Nations (ASEAN) committee on food, agriculture, and forestry. However, the field activities initiated and implemented by the SCSP ceased after 1983.

Accordingly, while SCSP has accomplished those objectives determined by its committee, other permanent regional bodies formed during its implementation period (SEAFDEC and ASEAN) will have the continued responsibility for coordinating regional development actions. This may seem to be a natural sequence for regional programs sponsored by UNDP and supported by inter-regional funds. Clearly, these regional bodies will eventually be entrusted with the responsibility and technical functions for developing and implementing different sectoral activities with additional technical support from bilateral aid. The important functions of FAO activities through the Indo-Pacific Fisheries Council will also continue. Bilateral aid programs have also provided new support to national programs, though regional management programs may seem to have a special need.

The problem of continued technical services therefore arises. Bilateral aid and the national programs tend to receive funding for limited periods only. The management problems of shared fisheries require permanent regional collaboration and technical support for the regional bodies that have been established, namely ASEAN and SEAFDEC. However, these bodies do not necessarily include all the countries of the region and are not sufficiently well funded to recruit the highest needed level of international expertise. Therefore the evolution that must eventually take place is for national expertise to undertake the various programs needed for efficient management of fisheries of the region. While there may initially be a problem in obtaining adequate technical personnel to work on such fishery programs, eventually it has to be accepted that objective independent technical management must develop, regardless of national policies. National policies should in fact be realistically determined on the basis of the technical expertise that is provided for the management and development of these shared fisheries resources. Though there may be political exigencies to accommodate national development schemes, yet in regard to the management of the regional resource of the shared-common-property fisheries, it is essential that the best technical expertise should objectively be devoted to carrying out this management efficiently. Failure to do so may see the consequent decline of available resources and costly subsequent efforts at rescue operations. Intensively fished areas of the world have not in the past shown a ready recovery to rescue measures.

Clearly, management for the maintenance of profitable sustainable yields of the resource is in the best interests of all. Objective technical capability to manage this should therefore be deliberately established during a regional program. Despite the initial difficulty of recruiting or training national personnel, through liaison with international institutions and agencies, the establishment of regional lead centres with appropriate staff in different disciplines is definitely necessary. The analysis and timely application of available data would probably continue being coordinated through the FAO regional fishery bodies. Thus the development thrust of the UNDP-funded program would continue to evolve and be ultimately carried out by nationals of the region with all required efficiency.

Accordingly, with an inter-regional program like SCSP, which has made such an effective start, as funding will not be available indefinitely, advance institutional plans for such monitoring, evaluation, and training functions must be made. In other regions of the world, regional programs may have shorter periods for coordinating inputs. SCSP was said to be most effective in gathering support for training programs and studies. If this can be considered a pattern for other regions, it seems essential that such regional programs should deliberately seek, within a given time, to establish lead centres able to pursue various technical disciplines and function in collaboration with other countries who are specially interested. This would apply not merely to the fishery resource management, but to technology of product development, training in aquaculture, and also various applicable fishing techniques.

It is worth noting that the goal here is not national superiority and preponderance of fish landings, but rather the profitably sustained yield of the resource that is shared by countries of the region. Both industrial high-seas fisheries and coastal artisanal fisheries are recognized as being significantly interdependent and essential for the regional economy. The ultimate goal is food and employment for people of the countries concerned, particularly in the rural or artisanal sector, and the effective education of people who operate small-scale artisanal fisheries. Food and employment for these people or consequent transmigration of populations into other sectors, are complex political problems. The subsistence fisheries and artisanal fisheries present a challenge throughout the developing world. If in this region solutions can be found, through coordination of policies and strategy for this most populous sector, then it is important that these solutions be carefully assisted and monitored to ensure success, as a pattern for development in other developing countries and regions of the world.

Finally, it may be significant to consider the suitability of bilateral inputs under the auspices of an inter-regional UNDP program. These could particularly involve activities in which there is acknowledged national expertise, and in which data derived from them can be objectively analysed under UNDP-FAO auspices, for sharing among all participant developing countries. Such inputs will still be identifiably bilateral, but will enhance and augment a UNDP program where limited funds are available.

Guidelines for successful project management

Preproject

1. Government must make an irrevocable commitment to long-range fisheries development, the project being part of an overall plan, even though donor assistance may only last, say, five years. This may be demonstrated by the establishment of a special project account deposited by recipient and donor agency.
2. Through stock assessment surveys, the project management should have a thorough knowledge of the fisheries resource: its availability, seasonal or cyclical performance, and its gross abundance for safe harvesting levels.
3. There must be established policies for the management of, and investment in, fisheries. These include regulatory measures, the

control of units and zones, data retrieval, and avoidance of the waste that results from overinvestment. (There must also be a clear government policy regarding production equipment at minimal costs in relation to market prices and fish imports.)

4. The project leader must be an experienced and responsible national with at least 10 years relevant experience in administrative, business, fishing, or technical operations. In liaison with government agencies he should assume responsibility for managing all project operation, and supervising subordinate staff. He should be assisted in technical decisions by expatriate advisers and by an advisory and coordinating committee. All staff for executive posts in projects should be impartially recruited, as being the best-qualified candidates that have requisite educational levels and aptitudes for dealing with fishing operatives or for serving in hardship locations. They should receive inservice training while serving as counterparts to expatriates and subsequently receive further training after proving their suitability in the program and interest in a career in such work. Salary levels are important.
5. The keys to successful management are:
   the selection of suitable leaders for the various aspects of the project and effective coordination of their activities;
   forward planning to prepare for operational needs and to avoid surprises;
   contingency planning to cope with emergencies and contain problems before they become crises;
   effective public relations and communication between fishermen, the public, and the project staff;
   and prompt response to public concerns.
   Finally, the activities that involve the greatest risk or uncertainty should be clearly identified and efforts made to minimize the risks.

During project

6. Before and during the project, national operational and management staffs should receive continual training on the job or overseas, to upgrade required skills. This needs adequate funds, objective selection of the best candidates with proven aptitudes, phased training at suitable institutions, a commitment for a stated period to post-training service with the project, and special arrangements for management staff in collaboration with regional, multilateral, or bilateral fisheries projects.

7. An advisory and coordinating committee (ACC) will facilitate the activities of the project and overcome difficulties. All groups that interact (such as port authorities, surveillance and technical education staffs, local government, university, and tourist operators) should be represented; *ad hoc* subcommittees should deal with such disciplines as marketing, pricing policies, loans, and tax incentives.
8. A monitoring and evaluation group (MEG) that includes, for instance, an economist, a fishing business operator, university researchers, and a development ministry planner, should independently review the results of pilot-level operations and suggest to project management and the ACC modifications that will guide current and future activities.
9. Operational flexibility is essential. Management should have the authority to act promptly when necessary at the project site. A small imprest account and supplies of gear and fuel will enable management to take advantage of favourable opportunities. Responsibility should be delegated to field operations as far as possible. Communication with headquarters by radio or radiotelephone is essential. The project activities must be responsive to changes in weather, seasonal migration, economic factors, labour or production costs, accidents or staff losses. The project should generate a wide range of products and possible applications to cope with changing seasonal or trade conditions for year-round stabilization insofar as possible.
10. Project activities are better concentrated in a small area until success is achieved, before dispersing staff and replicating the effort at other sites. The pilot effort should become a model whose limitations have been fully analysed and factually understood. Subsequent operations at other sites should proceed cautiously with reference to the experiences of the initial operation, making modifications for local fishing conditions and accommodating site-specific social or other operating circumstances. It may take 10 years to implant a permanently successful program, totally self-managed by national staff.

Postproject

The MEG and ACC should make clear recommendations for further developments so as to sustain the momentum of project activities when external inputs cease. These include periodic review seminars, post-project impact studies, and private sector takeover of certain operations, improved extension services, a refresher course training for field managers, advance planning for equipment replacement, etc. The continuity, take-up by the private sector and future role of government and project staff must be well planned and phased for lasting impact.

Opportunity and challenge

It has been seen that there is great potential for fishery development but considerable complexity in the management of factors required for the efficient harvesting of the resources and the establishment of successful fishing industries. There are many challenging ramifications of fisheries projects because of their multidisciplinary nature, and it may require the establishment of many precedents to improve the output of the industry, quality of fish products, and living standards of communities, while providing more and better fish as food. Fisheries projects seek to address all these social objectives and yet ensure profitability and continuity for the renewable resource. Success requires confidence, empathy, patient determination, and dedication from the policy-makers, fishery administrations, and development practitioners, in satisfying consumers and investors during the strategic modernization of fisheries.

Fishing is one of the oldest professions, harvesting the world's most elusive natural, renewable resource. The historic problems of fisheries management and development now are more likely to be resolved through coordinated, scientific, and strategic modernization in contemporary development programs. Despite the severe crises confronting the developing world, these programs can set the future pattern for efficient, sustained harvesting of the bounty of the planet's waters for Man's food, welfare, and survival, since water is the main source of life.

With a long past history, fishing valiantly serves the present generations but can contribute more to future populations through pragmatic experiences of investment.

Other books published by **Fishing News Books Ltd**

Free catalogue available on request

Advances in aquaculture
Advances in fish science and technology
Aquaculture practices in Taiwan
Atlantic salmon: its future
Better angling with simple science
British freshwater fishes
Business management in fisheries and aquaculture
Commercial fishing methods
Control of fish quality
Culture of bivalve molluscs
Echo sounding and sonar for fishing
The edible crab and its fishery in British waters
Eel capture, culture, processing and marketing
Eel culture
Engineering, economics and fisheries management
European inland water fish: a multilingual catalogue
FAO catalogue of fishing gear designs
FAO catalogue of small scale fishing gear
FAO investigates ferro-cement fishing craft
Fibre ropes for fishing gear
Fish and shellfish farming in coastal waters
Fish catching methods of the world
Fisheries of Australia
Fisheries oceanography and ecology
Fisheries sonar
Fishermen's handbook
Fishing boats and their equipment
Fishing boats of the world 1
Fishing boats of the world 2
Fishing boats of the world 3
The fishing cadet's handbook
Fishing ports and markets
Fishing with electricity
Fishing with light
Freezing and irradiation of fish
Freshwater fisheries management
Glossary of UK fishing gear terms
Handbook of trout and salmon diseases
Handy medical guide for seafarers
How to make and set nets
Introduction to fishery by-products
The lemon sole
A living from lobsters
Making and managing a trout lake
Marine fisheries ecosystem
Marine pollution and sea life
Marketing in fisheries and aquaculture
Mending of fishing nets
Modern deep sea trawling gear
Modern fishing gear of the world 1
Modern fishing gear of the world 2
Modern fishing gear of the world 3
More Scottish fishing craft and their work
Multilingual dictionary of fish and fish products
Navigation primer for fishermen
Netting materials for fishing gear
Pair trawling and pair seining
Pelagic and semi-pelagic trawling gear
Penaeid shrimps – their biology and management

Planning of aquaculture development
Power transmission and automation for ships and submersibles
Refrigeration on fishing vessels
Salmon and trout farming in Norway
Salmon fisheries of Scotland
Scallop and queen fisheries in the British Isles
Scallops and the diver-fisherman
Seine fishing
Squid jigging from small boats
Stability and trim of fishing vessels
The stern trawler
Study of the sea
Textbook of fish culture
Training fishermen at sea
Trends in fish utilization
Trout farming manual
Tuna: distribution and migration
Tuna fishing with pole and line